Foreword

Floods and other natural hazards are an ideal topic for the geographer, involving both physical processes and human responses. However, whilst a geographical treatment of flooding provides an important general survey, the specialists on flood problems are mainly from the engineering profession. Although students of physical geography are now contributing to the theoretical side of hydrology (see Weyman, 1975) their contribution on the practical side remains rather restricted.

The topic of flood hazard and flood protection is also the realm of economists, planners, and politicians. Thus advances in physical hydrology may take some time to be incorporated in the work of engineers and the best-developed schemes of engineers may be delayed by lack of funds, objections, and even parliamentary procedures.

There are basically two themes to engineering hydrology: water under under control (resource management) and water out of control (floods and droughts). Although this book has a floods theme, the problems caused by rapid temporal variations of river flow are as much a concern for water-resource management in Britain as those resulting from the geographical inequalities of runoff. Flood waters are not only damaging to life and property—they are also a largely wasted resource. The important role of flooding as part of a meteorgological/climatic/physiographic natural system is also stressed in the book where appropriate. The natural-systems approach also reminds us that conditions alter: climatic change may bring us more or fewer floods or even natural hazards of a different kind. It is surely a crucial test of scientific method to be able to research the problem until we are in a position to understand the system and manage it, or at least be aware of man's own influences on it.

The author would like to thank his former colleagues in the United Kingdom Flood Study team for inspiring his interest in the topic of this book and in particular Dr. J. V. Sutcliffe and Mr. M. J. Lowing for giving helpful advice. Considerable help with typing was provided by Mrs. Sue Hill and Miss Barbara Glover devoted much painstaking care to the illustrations. Drs. D. M. Harding, E. C. Barrett, and G. E. Hollis contributed offprints, Mr. F. Greenhalgh details of the Bath flood protection scheme.

Llwyn-y-Gog, May 1974   M. D. Newson

Contents

1 A brief review of documented floods in the United Kingdom

It would be difficult to define flooding rigidly in hydrological terms since the flood hydrograph is a basic property of rivers, resulting from the episodic nature of rainfall. The peak values of river discharge reached by the hydrograph form a data series from which floods above definitive thresholds can be chosen—for instance the bankfull discharge, which is a suitable lower limit for hazardous floods since man's activities on river flood plains are threatened. The mean annual flood is a value taken as the mean of a theoretical distribution applied to yearly flood peaks; it is this value which applies most to the engineer, particularly if he is equipped with the necessary conversions between the mean annual flood and much bigger events, for example the peak discharge of a 100-year return period. This chapter is mainly concerned with those floods which because of their intensity, duration, or destructive force have been recorded in some detail in the British Isles, irrespective of return period. Whilst Table 1.1 is clearly chronological there are certain regional regularities which are stressed here; West Country flooding has frequently been severe this century in response to record rainfall totals whilst north-east Scotland seems similarly afflicted. Both these areas are mainly upland with small, steep catchments and so the major floods to affect lowland, south, and east England are also described. The compendium of facts on these severe floods cannot easily be rationalized and in each case it is the missing data which are most important; the Forest of Bowland flood, however, demonstrates that, even when a hydrological instrument network is present to record data, the flood may so damage it as to necessitate the normal detective work used in ungauged situations.

Almost every flood is the subject of some sort of documentary record. The occurrence of heavy rain and of rivers spilling over their banks is usually marked by local newspaper coverage. Both the Meteorological Office and the Institute of Hydrology have news-agency schemes in order to receive information of unusual rainfall and river flooding—it is often some time before a flood event becomes quantified in terms of the data returned from observers in the area. Any events which cause damage may attain national news coverage and are often the topic of unpublished reports by river authorities, local councils, etc. Rarer floods are commonly the topic for scientific papers like the majority of those shown in Table 1.1.

TABLE 1.1

A sample of documented floods in the United Kingdom since 1829

Date of flood	Area/rivers affected	Author of paper	Journal/Publishers
Aug. 1829	Moray	Sir Thomas Dick Lauder	Black, Edinburgh, 1830
Jan. 1841	[S] Wiltshire (R. Till)	D. E. Cross	*Weather*, 1967
*Nov. 1894	Thames	G. J. Symons and G. Chatterton	*Qu. Jnl. Roy. Met. Soc.*, 1895
*Nov. 1894	W. Midlands/SW. England	A. Southall	*Qu. Jnl. Roy. Met. Soc.*, 1895
Feb. 1897	Bath (R. Avon)	Reporters/correspondents	*Keenes Bath Journal*
Aug. 1912	Norwich and E. Anglia.	Anon	*British Rainfall*, 1912
Aug. 1912	Norwich and E. Anglia	R. E. Pestell	*Norfolk Fair*, 1970
June 1917	Bruton (Somerset)	Anon	*British Rainfall*, 1917
Aug. 1917	N. Dartmoor	R. Hansford	*Trans. Dev. Ass*, 1930
May 1920	Louth (Lincs.)	P. M. Crosthwaite	*Trans. W. Engn.*, 1921
Aug. 1924	N. Somerset.	J. Glasspoole	*British Rainfall*, 1924
June 1930	Stainmore (Westmoreland)	F. Huddleston	*British Rainfall*, 1930
July 1930	NE. Yorks. (R. Esk)	English *et al*	*Whitby Lit./Phil. Soc.*
June 1931	Bootle (Cumberland)	Anon	*British Rainfall*, 1931
Nov. 1931	S. Wales	Pontypool R.D.C.	*The Griffin Press*, 1935
May 1936	Chilterns (Bucks.)	K. P. Oakley	*Records of Bucks.*, 1945
Jan. 1939	Ipswich (R. Gipping)	Reporters	*East Anglian Daily Times*, 1939
May 1944	Central/W. Wales	A. A. Miller	*Weather*, 1951
Mar. 1947	[S] Midlands and South	B. Howarth *et al.*	*Jnl. Inst. W. Engrs.*, 1948
Nov. 1947	[S] Inverness-shire	P. O. Wolf	*Jnl. Inst. W. Engrs.*, 1952
Aug. 1948	SE. Scotland	G. Baxter	*Jnl. Inst. W. Engrs.*, 1949
Aug. 1948	SE. Scotland	A. T. A. Learmonth	*Scottish Geog. Mag.*, 1950
Aug. 1948	SE. Scotland	J. Glaspoole and C. K. M. Douglas	*Met. Mag.*, 1949
Aug. 1948	SE. Scotland	A. Scott	*Scottish Agriculture*, 1950
Aug. 1952	Lynmouth (N. Devon)	C. H. Dobbie and P. O. Wolf	*Proc. Inst. Civ. Eng.*, 1953
Aug. 1952	Lynmouth (N. Devon)	G. W. Green	*Bull. Geol. Surv.*, 1955
Aug. 1952	Lynmouth (N. Devon)	P. Browne	*Readers' Digest*, 1972
Aug. 1952	Lynmouth (N. Devon)	C. Kidson and J. Gifford	*Geography*, 1953
Aug. 1952	Lynmouth (N. Devon)	A. Bleasdale and C. K. M. Douglas	*Met. Mag.*, 1952

Aug. 1952	Lynmouth (N. Devon)	W. N. McClean	Jnl Inst. W. Engrs, 1953
May 1953	Lochaber, Appin, Benderlock	R. Common	Scottish Geog. Mag., 1954
Aug. 1954	S. Wales	G. M. Howe	Met. Mag., 1955
July 1955	Weymouth	W. J. Arkell	Proc. Dorset Nat. Hist./Arch. Soc., 1955
July 1955	Weymouth	D. J. Paxman	Proc. Dorset Nat. Hist./Arch. Soc., 1955
Aug. 1956	Moray	F. H. W. Green	Scottish Geog. Mag., 1958
Aug. 1956	Cairngorms	P. D. Baird and W. V. Lewis	Scottish Geog. Mag., 1958
Oct. 1956	Border areas	R. Common	Scottish Geog. Mag., 1958
Aug. 1957	W. Derbyshire	F. A. Barnes and H. R. Potter	E. Midland Geogr., 1958
*Dec. 1960	Exeter	J. Brierley	Proc. Inst. Civ. Eng., 1964 (Discussion) 1965
*Dec. 1960	West Country and S. Wales	Surface Water Survey	Jnl Inst. W. Eng., 1961
*Dec. 1960	Exmouth	A. J. M. Harrison	The Surveyor, 1961
*Dec. 1960	S. Wales	G. Mcleod	Inst. Civ. Eng. 1970
Nov/Dec. 1965	S. Wales	G. Mcleod	Inst. Civ. Eng. 1970
Dec. 1966	Glan and Shin (W. highlands of Scotland)	G. Reynolds	Weather, 1967
June 1967	W. Wales	N. Rutter and J. A. Taylor	Weather, 1968
July 1967	Oxford	D. McFarlane and C. G. Smith	Met. Mag., 1968
Aug. 1967	Forest of Bowland (Lancs.)	J. A. Duckworth	Ass. Riv. Auths. Yr.Bk., 1969
Aug. 1967	Forest of Bowland (Lancs.)	F. Law	Fylde Water Board, 1968
*1967–8	Vale of York	J. Radley and C. Sims	Ebor Press, 1971
July 1968	Bristol N. Somerset	P. R. S. Salter	Met. Mag., 1968
July 1968	Bristol N. Somerset	P. R. S. Salter	Met. Mag., 1969
July 1968	Bristol N. Somerset	J. D. Hanwell and M. D. Newson	Occ. Pub. Wessex. C.C., 1970
Sept. 1968	SE. England	A. Bleasdale	Jnl. Inst. W. Eng.
Sept. 1968	SE. England (Esher)	Anon	The Engineer, 1968
Aug. 1970	Glenburn (N. Ireland)	R. Common	Irish Geographer, 1971
Aug. 1970	Moray	F. H. W. Green	Scottish Geog. Mag., 1971
July 1973	Central N. England	P. A. Smithson	Weather, 1974
Aug. 1973	Mid-Wales	M. D. Newson	Institute of Hydrology, 1974

* indicates that the paper contains a list of historical floods at the site.
[S] indicates snowmelt involved.

4   Flooding and flood hazard

Meteorologists tend to dominate the literature and the regular publication of *British Rainfall* from 1860 onwards has provided a vehicle for descriptions and analyses of each year's heaviest rainfalls. Descriptions of flooding are included but analysis of river flows is not within the journal's scope. Another important function of *British Rainfall* is to record the heaviest-ever falls in tabular form. The synoptic conditions producing the falls are described as part of the yearly reports. The three other major meteorological journals in Britain also feature in the Table. By contrast, the official yearbook of river flows, the *Surface Water Yearbook*, does not contain analyses of the effects of floods on catchments, most the hydrological aspects of flooding being covered by papers in the *Journal of the Institution of Water Engineers.* The frequency of flood papers from Scotland increased in the 1950s and 60s largely as a result of the scheme of flood surveys organized by the Royal Scottish Geographical Society.

The Lynmouth flood of August 1952 produced more important papers than any other flood. Two of them deal with the geomorphological effects of the flood, a trend which has been continued as part of the efforts of quantitative geomorphology to gather data on the magnitude and frequency of geomorphological action. Certain of the papers, as well as describing specific flood events, list historic floods at the site in question. Several other works provide such lists, for example the book *British Floods and Droughts* by Brooks and Glasspoole (Benn, London, 1928), in Northumberland and Durham by an anonymous book entitled *Memorials of the Floods in the Rivers of Northumberland & Durham* (1849), and the list provided for the Yorkshire Swale by Hugh Bowen Williams in a Ph.D thesis (1957).

August is the dominant flood month, although it is possible that if we base this assessment on the literature the situation is influenced by the intensity and suddenness of summer flooding which give meteorologists and hydrologists more material for producing reports. Another notable trend in the Table is the reappearance of certain areas on several occasions, for example the Moray area of Scotland in 1829, 1956, and 1970 (all August floods) and north Somerset and Devon in 1917, 1924, 1952, and 1968. The reason for such reappearances is partly physiographic, in that all upland areas produce a flashy runoff regime, and partly meteorological in that occurrences of heavy rainstorms seem more common in certain areas (probably a combination of storm track and orographic effects).

Since the most disastrous flood which remains in many people's memory was that in 1952 in Lynmouth it is natural to begin here with West Country floods. We begin with a very severe rainstorm which, however, produced only locally damaging floods in June 1917, when a depression moving up the English Channel brought over 200 mm of rain to the town of Bruton, Somerset, and over the Quantock Hills. The storm was extensive enough to give an east-west belt of country some 160 km long and 16 km wide over 100 mm of rain. Most of the fall was concentrated between the hours of 2200 G.M.T. on 28 June and 0100 G.M.T. on the 29th. Our knowledge of the rainfall amounts in Bruton is mainly due to the meteorological recordings of schools there. Although one unfortunate boy spilled what may have been a record fall whilst emptying the gauge at 0900 G.M.T. on the 29th, a boy called Mann at Sexey's School, Bruton, made sufficiently detailed observations of the amounts of water in both jar and can to allow a meteorologist to assess a fall of 242·8 mm in the previous 24 hours. This remained the highest gauged daily (0900) point rainfall in Britain until 1955, although there have been higher estimated falls. Although Bruton was flooded, the fact that the heaviest rain fell on a watershed from which streams radiate rather than converge is said to have prevented disaster. The resource aspect of such a storm can be gauged from the fact that the total fall amounted to some two billion litres of water.

The Cannington storm of August 1924 again occurred over north Somerset, although the synoptic situation was one of a stagnant depression over the North Sea. The storm was more local (about 130 km^2) and of a thundery type including hailstones up to a diameter of 150 mm. Flooding was locally severe due to the formation and subsequent breaching of dams of stones, straw, and vegetation across the torrents.

On Friday 15 August 1952 a storm occurred further west of Cannington for which the term 'physiographic aggravation' has been justly used to explain the far greater severity of flooding, for the storm centre was over Exmoor at the head of the extremely steep, narrow valleys of the River Lyn. Antecedent conditions were moderately wet for August and this also helps to explain the disaster. Loss of life (9 men, 16 women, and 9 children) was mainly the result of the development of the town of Lynmouth as a holiday resort in a situation of both beauty and great danger on either side of the River Lyn where it meets the sea. Despite the records of severe floods in 1607 and 1769 the stream had been obstructed, constricted, and neglected in the rush to build up tourism. For instance it is calculated that the West Lyn bridge, whose arch became blocked, causing extensive flooding in part of the town, was only half as wide as necessary to conduct such high flows. 20,000 million litres fell over the 100 km^2 Lyn catchment (giving an areal average of 143·0 mm).

In order to allow redesign of the town's waterfront in a less dangerous

manner the Lyn catchment was visited shortly after the flood by a number of scientists and engineers whose purpose was to estimate the rainfall
and river flow of 15 August. As McClean (1953), bemoans there was no
river gauging on the Lyn prior to the flood, nor did the raingauge network include a gauge within the Lyn catchment. Consequently the job
of the hydrologist was, as it has been many times since, largely one of
detective work.

The compilation of an isohyetal map for the Lynmouth storm was
based on records outside the area and a 'bucket catch' at Simonsbath
(Fig. 1.1). As in 1917 and 1924, the storm centre produced over 225 mm
of rain in the 24 hours up to 0900 on 16 August. However the area was
larger than at Bruton or Cannington. The recording raingauges in the
area were some way outside the storm centre (at Chivenor and Wootton
Courtenay). Their records, together with eye-witness accounts, suggest
two intense spells of rain at around 1700 and 2100 G.M.T. on the
Friday night. Erosion scars and estimated river flows suggest a maximum
intensity of 250 mm an hour during these periods (possibly lasting for
10 minutes or so).

Dobbie and Wolf's estimates of flow are based on a careful survey of
the channel cross-section (see Chapter 2) and the slope of the water
surface during the flood as evidenced by 'trash' marks left on the banks.
The hydraulic roughness is also required to calculate stream flow
velocities and there is some discussion of a suitable value for the Lyn
and its tributaries (Dobbie and Wolf, 1953). The resulting figures were
checked by summation of the tributary estimates to give that for the

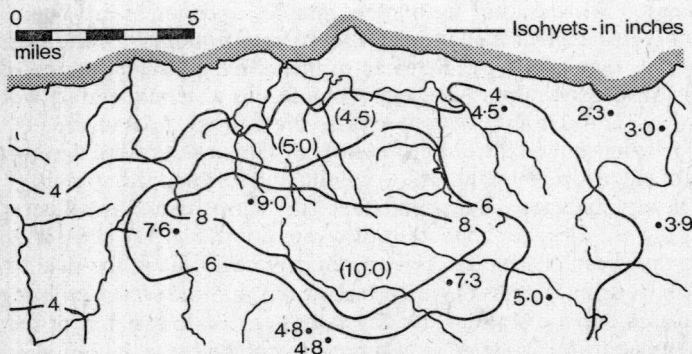

Fig. 1.1.   Rainfall over Exmoor and the Lyn catchment, 15 August 1952 (figures
in brackets are estimated) (from Dobbie and Wolf, 1953)

Lyn itself and by the construction of a 1:48 scale model (in paraffin wax at Imperial College) of a section of the Lyn's channel in order to define a river-level/river-discharge relationship. The resulting estimated discharge of 511 cumecs has only been exceeded on the Thames (drainage area nearly 100 times larger) twice since 1883. Although measures to reduce runoff are suggested, such as forestry (the reduction of extreme floods by forestry is regarded as insignificant in discussion), the sensible design of bridges and the careful maintenance of adequate channels are the chosen remedies. The adequate forecasting of such a flood in the future is not regarded as feasible although escape routes have been provided in Lynmouth. The rapid rise of the Lyn prevents adequate warning; for instance, during repair work at Lynmouth the river rose 3 m in 15 minutes. The discussion following Dobbie and Wolf's paper focuses mainly on the reliability of the peak flow estimate as a basis for design. Only one contributor ventures a hydrograph method capable of assessing the time parameters of flooding or of extrapolating the peak estimate to even more severe rainfalls.

In all the disaster cost £9 000 000 for loss of life, the destruction of 90 houses, 130 cars, and 19 boats, and the redesign of riverworks and roads. Twelve thousand gift parcels and £1 250 000 arrived as donations from seventeen countries for the stricken town.

The Lynmouth flood also received considerable attention from geomorphologists because of the vast deposits of boulders (estimates of up to 100 000 tonnes of them) left by the river and the extensive areas of slope failure on Exmoor (Green, 1955; Kidson and Gifford, 1953). It was estimated that the West Lyn dissipated 201 Megawatts of energy in its torrential flow through Lynmouth. Upstream super-critical velocities of 6—9 m per second were estimated and a boulder weighing 7½ tonnes was discovered in the basement of a hotel during cleaning-up operations. Green suggests that the boulder dump 0·8 km up the East Lyn which was not disturbed shows that a previous flood (possibly 1769) was even more severe. However, the prehistoric Tarr steps were displaced considerably for perhaps the first time in their long history.

As frequently occurs after such disasters, there were press rumours of dam collapse or cloud-seeding experiments as causes of the flood. The latter was never substanciated as a cause of the heavy rainfall and dam bursting was restricted to the collapse of a small embankment at Woolhanger. The most likely explanation of the sudden surges detected during the flood at Lynmouth was the formation and destruction of boulder and tree dams across the channel; these can be prevented in future by channel maintenance.

On 10 July 1968 flooding returned to north Somerset. The synoptic situation which produced the intense rainstorms during the night of 10/11 July has been described in general by Salter (1968, 1969) and in detail for the Mendip Hills by Hanwell and Newson (1970), who emphasize the chance combination of a variety of factors which produced the exceptional rainfall over Mendip; disruption of the circumpolar jet-stream, the sudden 'Jack-in-the-box' elevation of warm Mediterranean air behind a warm front associated with a Biscay depression, the presence of abundant atmospheric dust, and the drag exerted by the Hills themselves which allowed the following cold front to gain ground. Topographic funnelling by, for example, Cheddar Gorge which formed a 'wind lane' and the convection over the warmer south flank gave strong uplift over the Mendips. There was also heavy rain in the lee of the hills where low pressure developed whilst the air moved briskly over the relief barrier. Applying the Gumbel treatment (p. 17) to the fall of 130 mm at Bath, Rodda (1970) discovers a return period of between 10 000 and 100 000 years. However, a similar analysis applied to the discharge of two Mendip springs produces return periods of 15 to 40 years (Hanwell and Newson). Certainly there are newspaper records of severe flooding in Cheddar Gorge on 7 August 1931 and Burrington Combe was flooded in the 1890s. Since the Mendip plateau's shallow valleys are normally dry there were several cases of flood surges resulting from dams formed by stone walls across the valleys. This change from dry valley status to rushing torrent was the cause of most the geomorphological action resulting from the flood. The large amount of fill exposed both on the surface and in caves lends support to the fluvial explanations for some of the more dramatic karst features. Further evidence for the fluvial adjustment of the dry valley network comes from the agreement of predictions of the peak flow from dry valley drainage density and that estimated at Cheddar during the floods.

The floods caused severe damage in other parts of north Somerset, for example at Pensford and Keynsham where there were fatalities and in Bristol where the industrial area was hit. The effects of the flood on the population was said to have been long-lasting, being manifest in increased ill health and early mortalities. Heavy rainfall also occurred in a belt across the Midlands and into East Coast areas (Fig. 1.2).

West Country flooding of a different nature occurred in December 1960. The autumn and winter of that year were extremely wet, July to November being 57 per cent wetter than the 1916–50 average in the Exe catchment (Brierley, 1964). The reason appears to have been a more southerly track for depressions during that period; Scotland

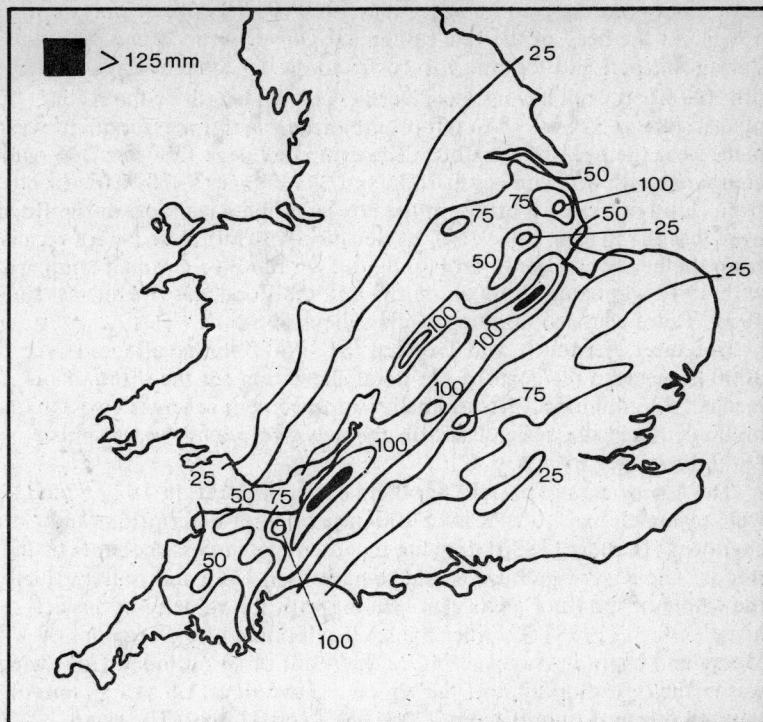

Fig. 1.2. Rainfall for 24 hours up to 0900 G.M.T. on 11 July 1968 (from Salter, 1969)

received 30 per cent less rainfall than average. The storm of 29/30 September and 5/6 October caused flooding in Exeter and Exmouth, the second flood producing almost double the peak of the first in Exmouth. Harrison (1961) uses an interesting indirect technique of discharge estimation near Withycombe Mill, Exmouth: velocity was estimated from the wave pattern around lampposts shown on photographs of the flooding. Fifty years of discharge records were then reconstructed from flood marks on the Mill. On this basis the design of channels and culverts to prevent future flooding was fixed around a return period of 74 years for the 6 October flood.

Further flooding occurred in Exeter on 20/1 and 25/6 October but the most extensive inundations were recorded at the end of this saturated period between the 2 and 5 December 1960 (*Surface Water Survey*, 1961). At the head of the Exe catchment Dulverton recorded 124 mm during this period. Flooding also occurred on the Avon at Bath on the 5th, the flood peak having travelled the 112-km length of the Avon mainstream in 31 hours. An important current meter measurement was made near the peak of the flood. The estimated peak flow was 365 cumecs compared with 367 cumecs in 1882 and 341 cumecs in 1894 (deduced from flood marks on a bridge in the city). On the River Severn the flood peak was the highest since 1947 at Bewdley which it took 2 days to reach from the headwater gauge on Plynlimon. On the Wye a similar comparison with 1947 was made although on the Usk the flood was the highest since 1931. There was also flooding in the valleys of South Wales.

In Exeter, Exmouth, and Tiverton the 1960 flooding affected over 3000 homes and the costs of the flood alleviation for the three towns reached £36 millions. Fifteen smaller improvement schemes cost £0·2 millions. Again the design basis for the new works was the December 1960 discharge.

The Moray area of eastern Scotland suffered floods in 1829 which were severe enough to warrant a 400-page book of descriptions and engravings (Lauder, 1830) drawing together eye-witness accounts of the floods. The River Findhorn is said to have risen 15 m and to have filled the whole of the floor of its glen 'passing with the velocity of a swift horse'. Green (1958) describes the severe flooding in the counties of Moray and Nairn in August 1956 as the result of an occluded front which was virtually stationary over the region for two days. Up to 250 mm of rain fell during 72 hours from 0900 G.M.T. on 31 July. The rivers Findhorn and Nairn were worst hit, although there was also flooding on the Spey. The Lossie came within a foot of the record August 1829 level in spite of the bold decision to let a new reservoir on one of its tributaries fill during the flood. In the same month there was flooding in the Cairngorms, with extensive mass movement on hillsides (Baird and Lewis, 1956) and over the Border areas of the Tweed basin (Common, 1956). August flooding badly affected the Tweed earlier in 1948, when 125 to 150 mm of rain fell over the Lammermuir Hills in one day (Glasspoole and Douglas 1949). Occluded fronts from a depression in the North Sea gave the Tweed over 20 per cent of its average annual rainfall and over £1 000 000 of damage was caused by flooding in Berwickshire.

Yet again, in August 1970 (Green, 1971), an occluded front covered Moray and Nairn giving a strikingly similar pattern of rainfall, confirming topographic effects. In this case, however, the 72-hour maximum was 150 mm, not 250 mm. As well as producing mass movement on the hillsides, the effect of such highland floods is to carry away the heavy bedload left by glacial deposition and leave it spread over fertile lowland farmland. As regarding Britain's larger rivers, the important years for extensive flooding in the last century have been 1894 (Thames, Wye, Severn, and Bristol Avon) and 1947.

The 1947 flood is commonly used as a rule-of-thumb comparative level for flood insurance and building planning in southern England. The levels reached were a considerable surprise to engineers who had not been taken unawares by severe widespread floods since 1894; it was the first severe flood to affect the River Boards. The winter was one of the snowiest on record and snow fell every day in some part of the British Isles from 27 January to 17 March. In most districts the ground was covered from 17 January to 13 March and a large area had a level depth of 300 mm. Easterly winds persisted and gave freezing conditions. Between 10 and 13 March a thaw spread north over England, in some areas accompanied by 25 mm of rain on the 14th and 15th (Howarth et al., 1948). The equivalent of over 100 mm of rain lay stored up for release by such a thaw and the frozen ground caused rapid runoff. The Rivers Lea, Great Ouse, Trent, and Medway were affected, according to published reports. Substantial areas of fen were drowned by the Great Ouse where flows were 50 per cent above the quantities formerly considered reasonably safe design maxima. Drastic remedies were planned, costing £4 200 000 in the Lea valley where urbanization had severely encroached on the floodplain below Hertford since the late nineteenth-century floods, £5 600 000 in the Fens, and £3 500 000 in Kent. However, analysis of the floods goes little beyond listing the peak flow at the few gauging stations and comparing rainfall (snow depth) with runoff. Upland storage reservoirs are suggested (and contested!) as a future flood remedy.

Cloudbursts have frequently affected Pennine valleys (such as those at Todmorden, Driffield, and Ikley in the late nineteenth century). That which affected the Forest of Bowland Upland on the afternoon (1630–1900 G.M.T.) of 8 August 1967 (Duckworth, 1969; Law, 1968) was the result of a narrow belt of intense thunderstorms moving eastwards across Lancashire in association with an unstable cold front. The Middle Knoll raingauge in the Dunsop Valley recorded 116·8 mm of rain in 90 minutes; all other raingauges (six) in the valley were flooded or buried by landslides. The Fylde Water Board's stream gauge on the river was destroyed by the rise of 5·8 m in 45 minutes. Three houses were destroyed in the village of Wray and ten more had to be demolished. The intensities of runoff are comparable with those at Lynmouth (discharges were estimated from debris, slopes, and roughness). Duckworth mentions almost £100 000 of essential channel improvements; he ascribes the severity of flooding to obstruction of channels by trees, bridges, culverts, and shoals of sediment, aided by the steep physiography of the area.

One of the most disastrous British floods of this century in terms of loss of life was at Louth, Lincolnshire, on 29 May 1920. The River Ludd has a 52-km^2 circular catchment on the chalk of the Lincolnshire Wolds, an unlikely physiographic situation for severe runoff. However, the rain of up to 153·4 mm (estimated from milk cans—116·6 mm recorded in the nearest raingauge) which fell between 1400 and 1715 G.M.T. gullied cultivated fields and produced rivers 18 to 30 m wide in normally dry valleys. The Ludd rose 5 m in 15 minutes and its constricted course between buildings at Louth was totally insufficient. By using the configuration of water levels up- and downstream of a bridge in the town an estimated flow of 152 cumecs was arrived at. The Ludd had never overflowed before and over £100 000 of damage was caused. The rainfall was produced from the north-east quadrant of a depression moving up the English Channel.

It was mentioned above that even a chalk catchment could not absorb all the runoff from very torrential rain in the case of the Louth flood (when frozen ground was certainly not involved). The Mendip limestone also produced rivers in normally dry valleys in July 1968 (see above). Another example of a chalk flood is reported by Cross (1967). The River Till in Wiltshire flooded five villages in January 1841, causing three deaths and £10 000 of damage after a warm front brought a rapid thaw and heavy rain. The Weymouth storm of July 1955 produced moderate floods but its greatest effect was on ground water recharge, boreholes recording rises of water levels of up to 12 m. The event was notable for the highest recorded catch in a rainfall day (275 mm) which was observed at Martinstown near Dorchester. Yet another example of runoff from dry valleys on chalk is provided by the very localized Chiltern storm of May 1936 (Oakley, 1945). The scarp topography and deep valleys are quoted as local triggers for this storm and parallels are drawn with the Yorkshire Wolds where such cloudbursts have been observed. There are also some similarities with the Mendip situation.

Dam bursts have been omitted from consideration here but it is worth recording the very great loss of life which accompanied the collapse of the Bilberry Dam, near Huddersfield, in February 1852 and the Bradfield Dam, near Sheffield, in March 1864. The Dolgarrog disaster (Snowdonia) in November 1925 led to the Reservoirs (Safety Provisions) Act of 1930.

2 The magnitude and frequency of river flooding from peak flow records

River-flow data

Rainfall data are more plentiful than streamflow data in Britain; the regular appearance of *British Rainfall* and the analytical activities of the Meteorological Office have ensured an efficient collection and publication of records. Whilst the *Surface Water Yearbook* lists summaries of river-flow data, including maximum mean daily flows for the year in question, it contains no analytical work. For floods of the same antiquity as known raingauge records we frequently have only flood marks on bridges to give us an indication of river levels. Whilst, in spite of its known disadvantages, rainfall is measured in a standard gauge, streamflow gauges are of various types depending on local physiography, data requirements and even budgetry constraints. Since the Water Resources Act 1963, which first commissioned the expansion of stream gauging, was more concerned with low flows than with floods, many streamflow stations simply do not measure floods.

Thus it is frequently the job of hydrologists to attempt an estimate of the peak flows based on indirect evidence such as eye-witness accounts or the debris lines left by the flood. Such methods have to be used in any case on the majority of flooded catchments, i.e. those which do not have a gauging station. The Lynmouth flood provides the best example of reconstructional estimation procedures. The basic 'slope/area' method of assessing the instantaneous peak discharge involves the use of one of the hydraulic flow formulae such as that of Manning (see Weyman, 1975). The method of survey for slope/area estimation is described in British Standard, BS 3680 (Past 5, 1970), and by Dalrymple (1956). There are several reasons for caution in using the method. Perhaps the greatest care needed is in the use of a roughness coefficient to represent the channel. There are three sources of information which can be used: tables constructed by experienced engineers (or from hydraulic experiments), photographs of channels with a specified roughness (see Barnes, 1967), or measurements of discharge made at the site during lower flows from which the value of n is established by reversing the equation. Where continuous river-level measurements are not made, the 'crest stage gauge' is useful for recording flood levels; it consists of a tube fixed vertically to the bank, inside which buoyant material such as charcoal rises with the river surface but adheres to the side of the tube when it falls. A pair of such gauges would give the gradient term in the Manning Equation.

The theoretical and empirical treatment of peak flows

As engineers began to collate documentary records of flood levels or install gauges to measure flow, their first attempts at general, rather than locally specific design rules came with a series of simple 'flood formulae', relating the maximum peak flow at a site to the catchment area (A), with a coefficient (C) and exponent (n) being fixed for the area in question, e.g.

$$Q_{max} = 825 \, A^{0.75} \quad \text{(India)}$$
$$Q_{max} = 3000 \, A^{0.5} \quad \text{(Scotland and Wales)}$$

(In both cases discharge is in cusecs and area in square miles)

The fastest advance in the development of flood formulae at the beginning of this century was in the field of storm-sewer design for urban areas. In 1906 Lloyd-Davies, following work in the United States by Kuichling (1889), gave British engineers the first of many versions of the *rational formula*. As well as catchment area (A) the rational formula uses a figure for rainfall intensity:

$$Q = C \, i \, A$$

where i is rainfall intensity and C a coefficient of runoff. Rainfall analyses such as that of Bilham (see Chapter Three) were largely motivated by the needs of sewer design.

The first body of design criteria for natural streams in use by British engineers was specifically for upland catchments on which dams were to be built and was contained in the Institution of Civil Engineers 1933 Report. Written primarily to aid the execution of the Reservoirs (Safety Provisions) Act of 1930, this Report was designed to describe the calculation of adequate sizes of spillway on earth embankment dams, thus preventing their destruction by overtopping. It was republished in 1960 with additional data from the intervening years, especially from the Lynmouth disaster.

'Upland' catchments are defined as not exceeding 101 km² and with a straight-line gradient from source to mouth of between 1 in 10 and 1 in 50. Although the storage term introduced by the passage of a flood wave through a reservoir is defined, the Committee on Floods in relation to Reservoir Practice, who wrote the Report, felt that insufficient evidence was available for storage 'when the stream or river traverses undulating or flat country'.

The rational formula is found to be of some use and new work is included on intensity/duration and intensity/ara relationships for rainfall by reference to the largest storms recorded to date. The rainfall intensity/duration formula is $I = 8/T + 1$ where I is intensity (inches per hour) and T is storm duration in hours. For this reason some further data are included in the 1960 edition. The method advised for calculating time of concentration is Bransby-Williams's formula (1922) which takes the following form:

$$\text{Period of concentration in hours} = \frac{L}{D} \times \sqrt[5]{\frac{M^2}{F}}$$

in which M is catchment area in square miles, F is the average slope in feet per 100 feet, D is the diameter of a circle of equal area to that of the catchment, and L is the straight-line distance from outflow to remotest watershed (miles). The Report chooses to standardize the value of F as 3·3 and L/D as 1·2 for upland catchments.

However, both the original Report and its revision are based mainly on recorded floods rather than theoretical methods. We are told 'The Committee has, therefore, been unable to put forward any rules for arriving at the probable maximum flood discharge by calculation from rainfall, and considers that it would only be safe to rely on actual records as showing what flood intensities have in fact occurred from catchment areas of different extent.'

The method of presentation for the data collected was a plot of runoff peak intensity in cusecs per 1000 acres of catchment against catchment area in thousands of acres (up to the 'upland' limit of 101 km² but also including other data up to 1818 km². Allard, Glasspoole, and Wolf (1960) extend it further to 4 444 km². They also graphically add the tabulated extra data in the 1960 reprint and replot all data on a logarithmic plot. Previously, on arithmetic axes the plotted data were enveloped by a curve enclosing the most intense floods. This curve related, therefore to 'normal maximum floods' and Fig. 2.1 shows how it was exceeded by estimates of flood intensity at Lynmouth. Rainfall data collected with the flood figures suggested two other situations; where rain falls for considerably longer than the time of concentration of the catchment its intensity is likely to be less and a 'prolonged catastrophic flood' results, whereas with violent short downpours a short rise to a very high flood (an 'acute catastrophic flood') will occur. The peak rates of acute catastrophic floods are described as being twice those of normal maximum floods yet these should still be passed with safety by an overflow weir built with a 2-foot

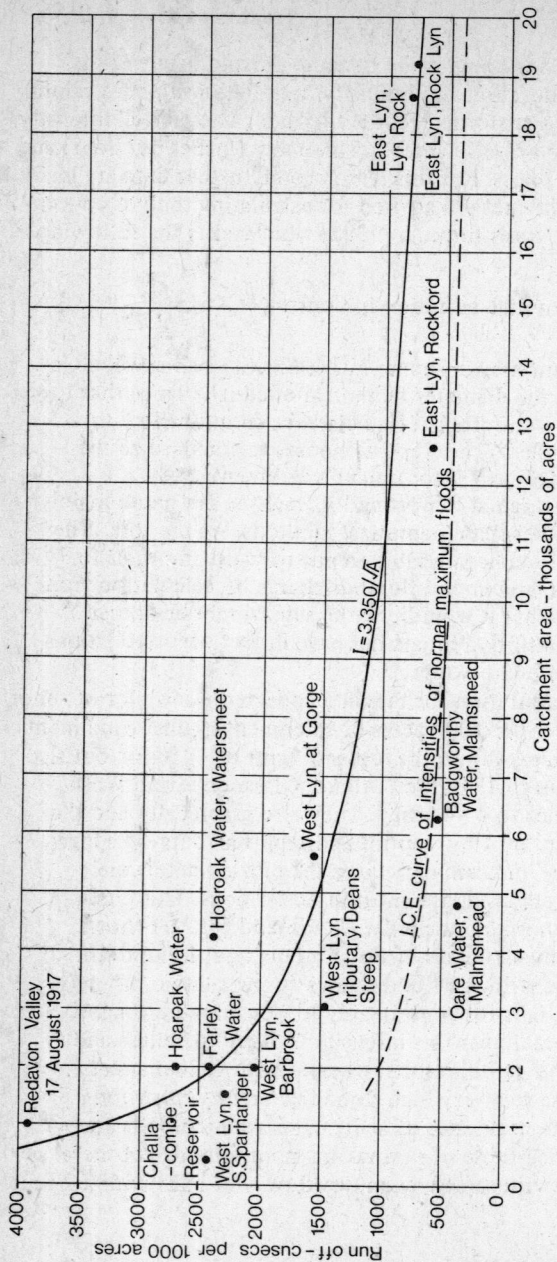

Fig. 2.1. The revision of the 'Normal Maximum' envelope curve following the Lynmouth disaster (from Dobbie and Wolf, 1953)

margin above normal maximum. From their list the Committee labelled normal maximum floods as being caused by rainfalls of about 3 inches in 24 hours, the bulk falling in the time concentration. The remainder of the Report deals with the approximate form of the normal maximum flood hydrograph and its lag through storage in those reservoirs larger than 2 per cent of their catchment area.

Chapman and Buchanan (1966) report than an investigation of recorded flood frequencies in the north and west of Britain has shown that the 'normal maximum flood' has no specific recurrence interval although the fact that values of 7 to 20 years were established is alarming. They recommend that 50 per cent be added to the 'normal maximum flood' suggested by the 1933 Report for substantial works although the basic 'normal maximum' figure may be used for minor river structures. Morgan (1966) assigns a 50-year return period to the normal maximum flood (calculated from floods over the last 50 years) and a 500-year return period for the catastrophic flood. He suggests that any design flood (for a return period, T) can be obtained from the catastrophic flood, multiplying it by $(T/500)^{\frac{1}{3}}$. Thus the factor for 500 years is 1·00, for 50 years 0·46, and for 5 years 0·22.

The design situation for dams requires an estimate of very high floods indeed since the dam should not be destroyed by natural hazard during its economic lifetime. On the other hand the effects of failure on a flood protection scheme or a sewer system are less drastic. For this reason the river engineer, municipal engineer, and planner require a treatment of the whole data series of flood peaks such as is provided by *magnitude/ frequency analyses*. This field of statistical analysis was largely open up by the pioneering work of E. J. Gumbel (1958) who treated the sample of peak flood discharges for any streamflow station (or drought flows at the same point) as a statistical distribution. Since both floods and droughts are extreme values from a data set with a full range of values Gumbel's work and attempts to refine it have concentrated on chosing a distribution which is appropriate to such extremes. Once chosen, the parameters of the distribution are fixed by reference to the available data sample and they can then be used to extrapolate the data to predict future extremes. Such statistical analysis treats any level of flooding as possible, in contrast to the meteorologists view that a certain maximum of precipitation may fall over a catchment and therefore provide a theoretical upper limit for floods. As Gumbel (1958) says, 'No physical law to account for these events exists . . . the relation between rainfall and runoff, far from being simple, is again of a statistical nature.'

The extrapolation of magnitude/frequency relationships based on fitting theoretical distributions to a sample of data places large strains on both the chosen distribution and the data used. Let us consider first the nature of *extreme-value distributions*. Perhaps best known in the analysis of floods is the Type I extreme-value distribution of Fisher and Tippett (1928) applied to hydrological data first by Gumbel (1941); Powell (1943) prepared the plotting paper based on this distribution. A further distribution advocated by the United States Water Resources Council (1967) is the Pearson Type III which has three parameters. Plotting paper is not available for this distribution.

Estimation of the parameters of the chosen distribution may be performed by the method of moments or by the principle of Maximum Likelihood. However, there are simplified tables to aid calculation (e.g. McGuiness and Brakenseik, 1964) and the graphical method of Gumbel/Powell is simpler still (see Fig. 2.2).

Using extreme value probability paper a *plotting position* against an appropriate recurrence interval is required for each flood discharge in the available sample. The formula for plotting position varies but Weibull's,

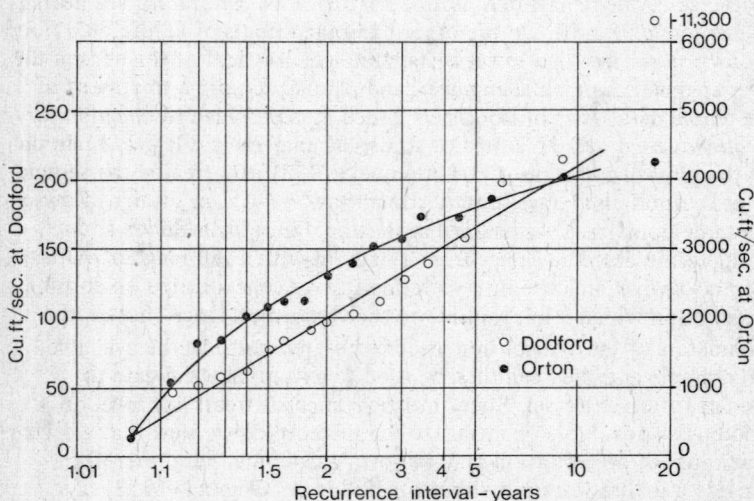

Fig. 2.2.   Flood frequency analysis for two stations on the Nene and Great Ouse using the Gumbel/Powell probability paper and a plotting formula of $(n + 1/m)$

chosen by the United States Geological Survey (Riggs, 1968), is

$$T = \frac{1}{p} = \frac{(n+1)}{m}$$

where T is recurrence interval in years, p is probability of exceedence in
any one year, n is the number of items in the sample, and m is the rank
order (from biggest to smallest) in the sample array. An earlier formula
by Hazen (1930) is sometimes used in which m is divided by $(2n-1)$;
further formulae have been devised to avoid the effects of bias in plotting.
If the resulting cluster of points describes a shallow curve, a fit by eye is
valid; only when a straight line is plotted through the points is strict
membership of the appropriate distribution assumed. Although graphical
fitting is simple and rapid it does not have the objectivity and repeat-
ability of mathematical methods. The latter are mostly performed by
computer and this allows calculation of extra information such as the
goodness of fit of the data to the chosen distribution (by the Chi-square
or Kolmogorov-Smirnov tests) and confidence limits for the parameter
estimates.

Wilson (1969, pp. 155–64) provides an example of various plots
from River Thames data and shows that the choice of probability paper
can be largely subjective. As for the reliability of the data, the quality
of each discharge value is of course important but even data of 100
per cent accuracy are unsuitable if only available over a short period.
Benson (1960) used a hypothetically perfect 1000-year record from
which to draw groups of various records for extreme value analysis
and listed the following lengths of record necessary to define the T-year
flood within 10 per cent of the correct value 95 or 80 per cent of the
time:

TABLE 2.1
*The effect of data length on estimates of flood recurrence intervals
(if available)*

Reccurrence interval *T*	*Length of record in years*	
	95% of the time	*80% of the time*
2·33	40	25
10	90	38
25	105	75
50	110	90
100	115	100

Source: Benson, 1960

Although the instantaneous maximum discharge for the year is commonly chosen as the basic item of data in flood frequency analyses (forming the *annual maximum series* where N years of data are available), one method of getting more information from recorded data is to tabulate and use all instantaneous peaks over a certain threshold (q_0), thereby generating a *'partial duration series'*. Perhaps two or more floods each year are then involved in the fitting of statistical parameters. However, Cunnane (1973) has demonstrated that for return periods of more than ten years the latter method has a larger sampling variance than the former. For smaller floods the partial duration series does yield more precise estimates (thereby justifying the extra work involved in producing it) if an average of 1·65 flood peaks per year are included (or more).

One of the most frequently chosen ways of obtaining a reliable data base from mainly short records is *regional analysis*. For instance Dury (1959) analysed data from the *Surface Water Yearbook* for stations on the Nene and Great Ouse. By using stations in the same region, a long common base-period for analysis can be obtained by generating missing values in short records from the correlation obtained between overlapping periods of these and a long-duration, good-quality base station. Once extreme value analysis has been performed on each of the new records, the choice of region is tested by assessing the homogeneity of the ratio of 10-year to 2·33-year floods (see Dury, 1959; Cole, 1966; Biswas and Fleming, 1966). If homogeneity is proved, the *mean annual flood* ($Q_{2.33}$) can be related to catchment characteristics such as area. Biswas and Fleming also plot a regional frequency curve of the median value of the ratios between the 10-, 20-, and 30-year floods to $Q_{2.33}$ for the stations used. Thus to obtain the 30-year flood for an ungauged basin one finds $Q_{2.33}$ from its catchment area and multiplies it by the factor suggested by the *regional frequency curve*.

Cole (1966) covers the whole of England and Wales and finds sufficient variation in the relationship of $Q_{2.33}$ with area to justify the mapping of six regions. A regional frequency curve for calculation of higher recurrence intervals was also compiled.

A more thorough way of investigating the regional variation of floods in relation to catchment area is to include in the analysis with the mean annual flood such other catchment characteristics as slope, drainage density, average rainfall, geology, soils, and other factors which are derived from maps and have a logical relationship with floods. The United States Geological Survey has led the way with this style of analysis through the work of Benson (1962, 1964). By using 164 catchments in New England, U.S.A., he obtained *multiple regression* expressions for $Q_{2.33}$ as the dependent variable with up to six independent map indices. In Britain Nash and Shaw (1966) and Rodda (1969) have also used multiple regression.

Rodda's equation employs *drainage density* as a variable but care is needed in the scale of map used and the definition of channel to be measured for drainage density. The simpler *'stream frequency'* which involves counting stream junctions in a basin and dividing by area has been used successfully in multiple regression analyses by the U.K. Floods Study. Thus the average value for mean annual floods countrywide is:

$$\bar{Q} = 0.0201 \text{ AREA}^{0.94} \text{ STMFRQ}^{0.27} \text{ S}1085^{0.23} \text{ SOIL}^{1.23} \text{ RSMD}^{1.03}$$
$$(1 + \text{LAKE})^{-0.85}$$

where STMFRQ = stream frequency, S1085 = a measure of main channel slope, SOIL = a measure of winter rain acceptance by soils, RSMD = a measure of runoff excess, and LAKE = the proportion of the catchment occupied by lakes or reservoirs. The equation is based on 530 streamflow records.

For six regions of the British Isles (Ireland is one region) the coefficient can be modified to give a more accurate prediction. A further regional subdivision is used to provide the multipliers for higher return periods. These multipliers for use with the mean annual flood range from 1·37 to 1·65 for the 10-year flood, from 1·96 to 3·56 for the 100-year flood, and from 2·14 to 4·46 for the 200-year flood.

What do the values of recurrence intervals mean to the engineer who has used an empirical flood study of the sort described above to determine the mean annual or higher recurrence interval flood at an ungauged site (or performed statistical analysis of records from a gauged catchment)? Most certainly they do not mean that works constructed to withstand the 100-year flood will not be submerged for the next century. As Herschfield and Kohler (1960) (see Fig. 2.3) point out, if the engineer wants to be 90 per cent sure that the works will not fail within 10 years of completion, they should be designed for the 100-year event. Gumbel (1955) advocates that it is the *calculated risk* and *desired lifetime* which should replace recurrence interval in design. At the moment, however, there is no sign of this and the two most popular recurrence intervals for design are between 1 and 10 years for urban drainage and around 100 years for flood protection schemes. However, it is quite usual in cases where a scheme is prompted by recent severe flooding to design around that flood discharge (e.g. Exeter, Bath, Lynmouth); this has considerable advantages for public relations. Any more sophisticated approach requires not only more hydrological data but many more economic data.

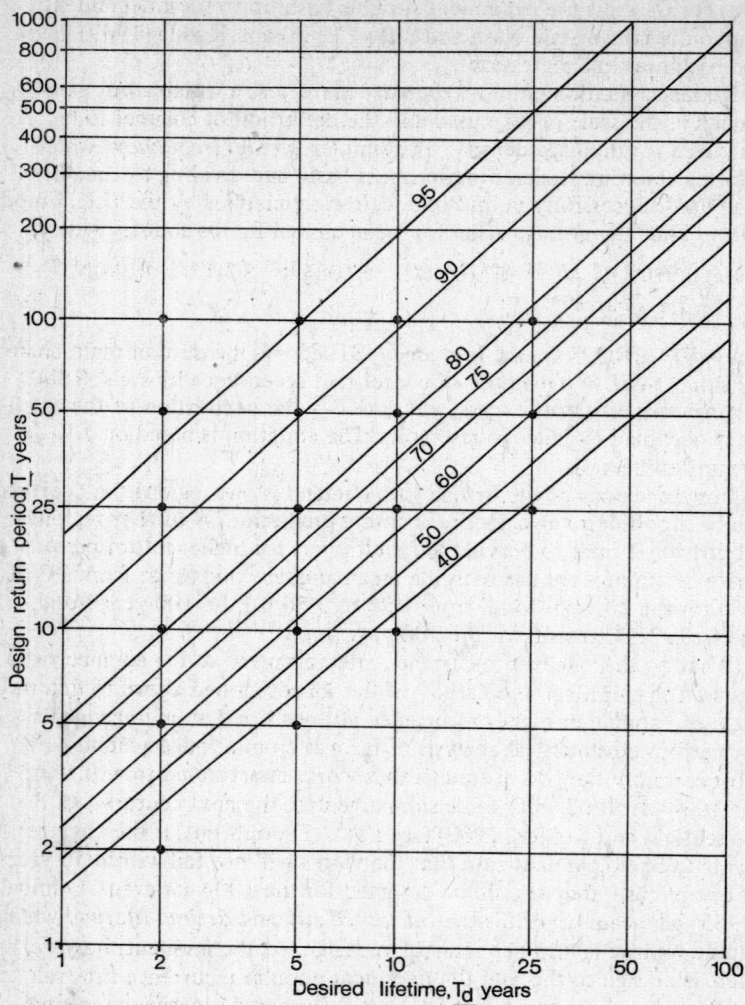

Fig. 2.3.   Calculated risk. The oblique lines represent observed probabilities of
not failing in *Td* years (from Herschfield and Kohler, 1960)

3 The magnitude and frequency of river flooding from rainfall data: the flood hydrograph

Rainfall data and analysis

Since heavy rainfall is the major cause of flooding and because rain gauging has both a longer duration and a denser network than river gauging in Britain, rainfall analyses are basic to flood analyses. There are over 6000 daily raingauges in the United Kingdom and possibly over 600 recording gauges. As well as the statistically analysing their records to predict the deepest, heaviest, longest spells of rain, we need to analyse the synoptic weather conditions leading to heavy rain; these are of fundamental interest for flood warning.

There are a number of geographical and seasonal regularities in heavy falls of rain. Bleasdale (1963) has summarized these for the period 1863 to 1960 as reported in the pages of *British Rainfall*. He finds that no part of the United Kingdom can be considered safe from daily falls of at least 100 mm. Thus be considers a list of 142 occasions on which 125 mm or more were recorded during the period considered. This list conceals two sources of error: the restriction of entries to the standard 0900–0900 G.M.T. 'rainfall day' (ending with the emptying of the 'day's' catch) and the ommission of the areal characteristics of the storms. Thus falls of 250 mm or so which were recorded during unrestricted 24-hour periods do not enter the list and, whilst the Cannington fall of 1924 ranks higher than the Norwich one of 1912, the area affected by the latter was vastly greater.

Furthermore we must also bear in mind the fundamental inaccuracies implicit in the design of a raingauge which may be exaggerated during heavy falls and that the geographical distribution of gauges tends to give the Western Highlands of Scotland fewer entries in the list than the Lake District where there are more gauges.

From his total of 142 occasions, 103 (over 70 per cent) belong to the five mountainous high-rainfall areas in the west: Dartmoor, South Wales, Snowdonia, the Lake District, and the Western Highlands of Scotland. Frontal and orographic influences are clearly at work here, combining to give such widespread mountain falls as in November 1931. Winter is the main season for such falls.

November and December are consequently likely months for floods,

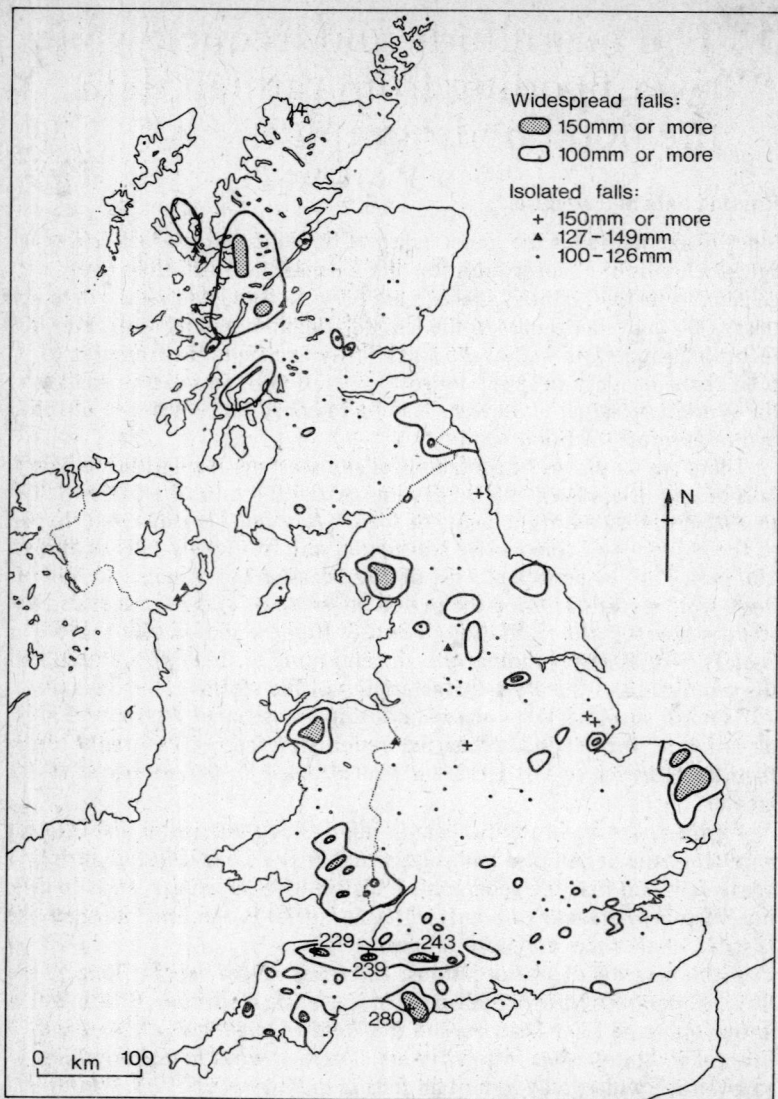

Fig. 3.1. Distribution of the largest daily rainfalls recorded in the United Kingdom 1863–1960 (from Rodda, 1970)

followed by August which is the main month for the convective, thundery storms which give the heaviest falls in the lower-rainfall easterly areas. The 39 occasions listed for such areas show a geographical clustering near the east coast but a strong localization of falls exceeding 225 mm in south-west England. Rodda (1970) maps the location of the heaviest falls in this period (Fig. 3.1).

Although Bleasdale's treatment ends in 1960 he updated his list in 1970 to include nine additional entries and bring the ratio of winter to summer falls up to almost exactly 3:1. In terms of the total volumes precipitated, the storms of July and September 1968 rank among the three or four largest, away from the mountains, this century.

A rather different analysis of heavy falls between 1956 and 1971 was performed by Finch (1972). Taking 22 gauges around the country he defines exceptional falls on the basis of river flooding, the threshold for which appeared to be 30 mm in two consecutive rainfall days. Finch uses the data to study three topics, increasing incidence of heavy falls, seasonal distributions for different regions, and synoptic types associated with the falls. Although he used few upland stations, his conclusions are similar to Bleasdale's on the seasonality of rainfall. The further a station is to the north-west the more important is the cold front wave for heavy falls. Warm fronts seldom produce exceptional falls although the slow-moving occlusion is important in southern England. Such results can be of value to the forecaster. Crossley and Lofthouse (1964) also show a south-east bias in severe thunderstorm activity. They map the incidence of such storms over a 20-year period for the months of May to September.

The probabilistic approach to heavy falls is taken by Rodda (1966, 1967, 1970). Using 121 stations, nearly all of which had 50 years of record, he too analysed the daily (0900 G.M.T.) totals, subjecting them to Gumbel Type 1 probability analysis. He was able to construct maps for various return periods of daily fall on the basis of a strong relationship between the one-day falls and the station annual average rain (Fig. 3.2).

Both Bleasdale and Rodda restrict their analyses to daily falls but engineers badly need information on storms as they occur, unrestricted in period and with hour-by-hour detail of rainfall intensities. Thus the recording raingauge is an important source of data although it is less widespread and with shorter records than the daily gauge. An expansion of the recording raingauge network took place partly in response to the Lynmouth flood, partly (in cities) during a drive to improve sewer design, and finally as part of the hydrometric network envisaged by the Water Resources Act.

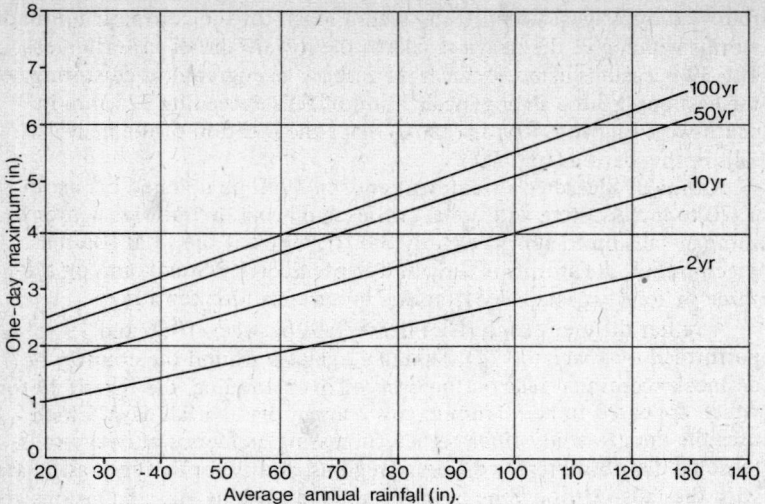

Fig. 3.2.   Relationship between average annual rainfall and one-day rainfall
amounts for different return periods (from Rodda, 1967)

The earliest attempt to study short-period intense rain (Bilham, 1935) is still the best known. Bilham's main aim was to refine, on a numerical basis, the ordinal classification of short-duration falls ('noteworthy', 'remarkable', and 'very rare') used hitherto in *British Rainfall*. The regularities he discovered did, however, yield an equation of great value to drainage and sewer authorities. Using 18 recording gauges with 10 years of data he selected 12 as being representative of the drier areas of the British Isles. The data consisted of the minimum durations of 5, 10, and 25 mm falls. The number of falls in the 10 years were grouped for each size of fall into those in 6 minutes or less, 15 minutes or less, 30 minutes or less, and 60 minutes or less. Bilham found that the number of occasions in 10 years (n) on which 5 mm fell in a certain duration (t) equalled $90t$; for 10 mm falls the relationship was $n = 15t$; and for 25 mm falls $n = 9t$. The general form of the relationship was

$$n = 1 \cdot 25t \, (r + 0 \cdot 1)^{-3 \cdot 55}$$

where r is the given rainfall in inches.

Bilham drew curves of the $n = 10$ (one day a year) to $n = \frac{1}{2}$ (one day in twenty years) rainfall durations and used these to fix the new definitions of 'noteworthy', 'remarkable', and 'very rare' falls (Fig. 3.3). He considered

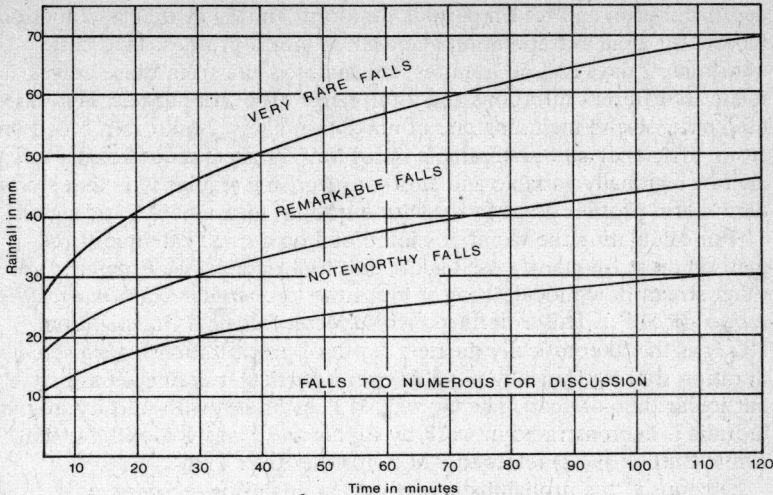

Fig. 3.3.   Classification of heavy rainfall (from Bilham, 1935)

that falls occurring in less than five minutes were beyong the accuracy and resolution of the existing recording gauges. However, such fine intervals are of importance and later several steps were taken to refine Bilham's work in the light of better gauges and denser networks. Holland (1967) describes the operation of more than 16 recorders at approximately 1 kilometre intervals at Cardington. Two-minute time steps are used for analysis of shower patterns, the reduction factor between point and areal falls, storm profiles, and storm drift. For 2-minute rainfalls the percentage reduction of a fall recorded at one gauge for extrapolation to the surrounding area is simply the square root of that area in hectares. For duration between 2 minutes and 3 hours a gamma function is involved and in a later publication devoted to updating Bilham's work Holland (1968) presents a graph of the reduction factor. The publication also tabulates the results of a Gumbel treatment of rainfall intensities, giving the 1, 2, 5, 10, 20, 50, and 100-year return-period intensities for durations of between 2 minutes and 24 hours. Despite this improvement, Bilham's formula is largely vindicated although the constant of 1·25 is changed to 1·39.

The Meteorological Office's contribution to the United Kingdom Flood Study used 600 long-term raingauges with an average of 60 years of record, 6000 for the last decade, and 200 autographic records to get depth/duration and return periods for storm rainfall as well as reduction factors for areal extrapolation and average storm profiles. Two basic durations, 2 days and 60 minutes, are analysed and from these curves relate to different durations and a full range of return periods. Maps are used extensively, including one of maximum likely 2-hour falls based on storm efficiency and precipitable moisture. The areal reduction factors are not regionally variable and another surprising regularity is that standard storm profiles do not vary with duration, return period, or region.

For calculating the worst possible flood on a small catchment the method most frequently used is the unit hydrograph (see Chapter 4) or other streamflow model, using as input the *probable maximum precipitation* (P.M.P.). This is defined (World Meteorological Organization, 1973) as the 'theoretically greatest depth of precipitation for a given duration that is physically possible over a particular drainage basin at a particular time of year'. The use of P.M.P. estimates with unit hydrographs in India is demonstrated in work by Binnie and Mansell-Moullin (1966), whilst Butler (1972) refers to P.M.P. for the River Trent.

Obviously, the probabilistic methods of analysing recorded peak flows described in Chapter 2 and those of analysing rainfall for use with unit hydrographs are complementary. The probability of the rainfall cannot, however, be directly applied to that of the flood because of the intervening probabilities of antecedent conditions and other factors. Nevertheless, estimates based on rainfall are considered more reliable for floods of high magnitudes, especially probable maxima.

The flood hydrograph

As well as determining the probability of occurrence of flood peaks of a selected magnitude, there are several situations in which engineers also attempt to predict the whole flood hydrograph. The extra information is essential:

(a) where, as in most cases, river-flow records are too short to provide reliable estimates of floods of a very long return period. In this case the greater length of most rainfall records can be used for the simulation of big floods by means of a hydrograph model and low-frequency rainfalls, such as the probable maximum precipitation;

(b) where some knowledge of the duration of flooding is required—for instance agriculturalists may require to know how long flood plains will be submerged;

(c) for real-time forecasting of floods using incoming information on rainfall over the catchment.

The *unit hydrograph* was first described by Sherman (1932). It consists of an average response function for the quick-return runoff of the catchment determined by analysis of storm rainfall and flow hydrographs for a number of flood events from the catchment for which design is required. By studying events from a number of catchments, the features of the unit hydrograph for the ungauged catchment (the more usual design situation) can be predicted by regression equations using catchment or storm indices as independent variables. The method is described and illustrated by Weyman (1975).

The first analytical stage is to separate the *quick-return runoff* or *storm-flow* from the *base-flow* which results from steady depletion of soil moisture and ground-water storages. In most cases those parts of a river-flow hydrograph occurring between flood peaks will give an indication of the slope of the *recession curve* of base-flow and this can be used for separation during flood events. Most authors use a separation procedure unique to their research interests ranging from sophisticated mathematical extrapolation to chemical sampling.

After the separation of the hydrograph the volume of quick-return runoff is planimetered and compared with the volume of rainfall. At this stage the runoff during each time interval (usually 1 hour) of the hydrograph's duration can be plotted as a ratio of the total runoff and expressed as a *distribution graph* with time as abscissa. The averaging of the ordinates of a series of distribution graphs for floods on the catchment gives an elementary hydrograph for design purposes.

The way in which the ratio of runoff to rainfall is expressed and the method subsequently used to separate the rainfall hyetograph into an *effective rainfall* volume (equivalent to the runoff volume) and '*losses*' are also points for conjecture in hydrograph analysis. For the sake of simplicity at this state the relationship of runoff to rainfall will be expressed as a percentage and this percentage of the total rainfall will be separated with a straight line horizontally crossing the hyetograph.

The next step isolates a fundamental parameter of the unit hydrograph and one which is arousing much research interest at this time. The time lag between rainfall and runoff can be expressed in a variety of ways. Perhaps the simplest for an uncomplicated hydrograph is *rise-time*, the duration of the rising limb of the hydrograph. Various expressions of *lag-time* involve the time between the first rainfall and peak flow, whereas most expressions of *time-to-peak* use the centroids of gross rainfall and separated runoff (where peaks are multiple). The various measures are highly correlated and, as with hydrograph separation, consistency is the most important constraint in analysis.

The simpler methods of unit-hydrograph derivation are inadequate to cope with the complex sequences of rainfall and flow hydrographs encountered in nature. To cover such complexities usually involves the solution of equations linking the rainfall and runoff of successive time increments until all the unknowns (the unit hydrograph ordinates) are solved. There is a trial-and-error solution which is used as an exercise by Wilson (1969, pp. 115–21). W. M. Snyder (1955) fits the coefficients defining the time distribution of runoff by least squares (matrix algebra is used).

Perhaps the most widely known British application of the unit-hydrograph principle is that of Nash. He assumed that the unit hydrograph's form could be specified before analysis by constructing a model consisting of a general equation with two parameters. The equation assumes that the operation performed by the catchment on an instantaneous rainfall (it is an *instantaneous unit-hydrograph* equation) is equivalent to that of a succession of linear storage reservoirs. The form of the equation allows the method of moments to be used in finding the best-fit values of the two parameters when the model is fitted to a complex flood (Nash, 1957, 1959). It is these parameters which are then related to independent variables (see below). Dooge (1959) prefers to lump the catchment reservoir representation and claims greater physical reality, if less convenience.

O'Donnell (1960) uses the analysis of the harmonic coefficients of the curves of excess rainfall and its resulting runoff hydrograph to yield the harmonic coefficients of the linking unit hydrograph. The runoff and rainfall graphs are viewed as sine-cosine Fourier series. Overton (1968) also uses this concept and fits the coefficients using Snyder's least-squares method.

The more sophisticated methods have become viable since the introduction of computers. Even with sophisticated techniques the resulting unit hydrograph is sometimes far less stable and ideal a shape than is required for further analysis when fitted to complex floods.

As mentioned above, the derived unit hydrograph for a catchment, or the average of several, can be used to obtain that catchment's peak discharges of long return period by convoluting with a storm rainfall input of similar rarity. Alternatively, comprehensive studies of unit hydrographs on a lot of catchments may be used to develop *synthetic unit hydrographs*, so called because the relationship of the index parameters (such as time-to-peak, peak discharge per unit area, and time-base) to catchment variables allows their synthesis for ungauged catchments. One of the first of such studies was that by F. F. Snyder

(1938) who hypothesized that the time lag between the rainfall centroid and peak flow depended on the time-distance from the outlet of the basin to the centroid of the time/area diagram (see Wilson, 1969, Figs. 8.16 and 8.17). Snyder's method is described more fully by Weyman (1975).

One of the most widely used synthetic hydrograph methods is that derived by the United States Soil Conservation Service (Reich, 1962). It can be used on watersheds up to 100 square miles in extent. The basic source for the method is that by the U.S.D.A. (1957) but Reich gives an ideal summary. A basic dimensionless hydrograph, representative of 'a large number' of natural hydrographs from all over the U.S.A. is approximated as a triangle. In this hydrograph the recession limb (T_r) is commonly found to occupy $\frac{5}{8}$ of the total hydrograph duration and the time-to-peak (Tp) $\frac{3}{8}$. If the peak discharge of the triangle (i.e. its height) is taken as qi, the total discharge, Q, by geometry is

$$Q = qi \cdot \frac{Tp + Tr}{2}$$

whence the peak rate

$$qi = \frac{2Q}{Tp + Tr} \text{ inches/hour}$$

Since $Tr = 1 \cdot 67\ Tp$ we may substitute to give:

$$qi = \frac{0 \cdot 75Q}{Tp} \text{ inches/hour}$$

Since 1 inch/hr./sq. mile produces 645 cusecs the equation becomes

$$qi = \frac{484 \times \text{Area} \times Q}{Tp}$$

Time-to-peak as defined in this chapter is equal to the catchment lag (L) plus half the rainfall duration ($D/2$). The lag has been found to be 0·6 times the time of concentration which is related to the catchment characteristics of length and relative altitude by means of a nomograph (Fig. 3.4).

The design rainfall amount and duration are then selected from local meteorological knowledge. The proportion of the rain which becomes runoff will clearly vary with soil type infiltration and plant cover, and tables are provided which allow the choice of a 'curve number' on a graph relating runoff to rainfall.

Having plotted qp against the time Tp, the triangle is completed by

L in feet

600 800 1,000 2,000 3,000 4,000 6,000 8,000 10,000 20,000 30,000 40,000

0·1 0·2 0·3 0·4 0·6 0·8 1·0 2 3 4 6 8 10

To in hours

Example

H in feet
(omit gully overfalls and waterfalls)

1000 800 600 400 300 200 100 80 60 40 30 20 10

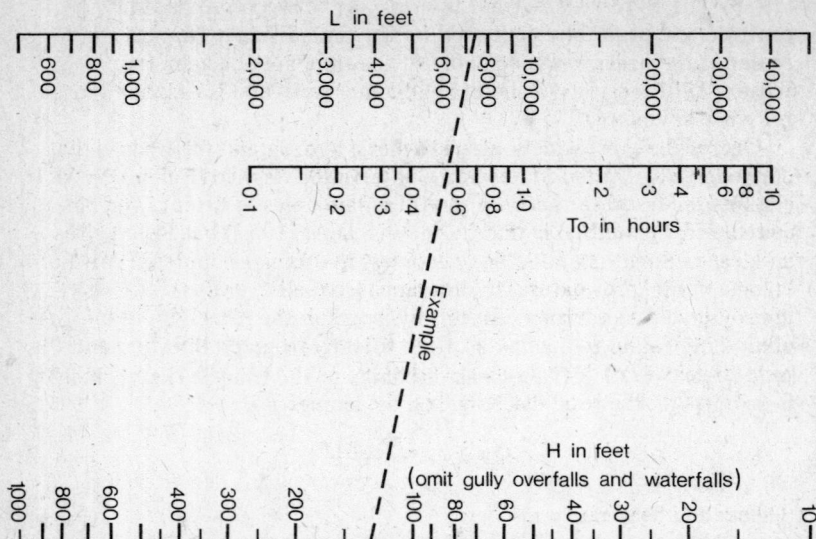

Fig. 3.4. Time-of-concentration nomograph (from Reich, 1962, based on U.S.D.A., 1957)

zeroing again at time $(Tp + Tr)$. The rough triangle can be smoothed to look more realistic using methods described by the U.S.D.A. If the design storm is a complex one each increment of rain can be used to generate a triangular hydrograph which is lagged by the correct amount, and summation of the ordinates gives the final hydrograph. However, if the rainfall excess occurs at a fairly uniform rate a single hydrograph will suffice.

A vindication of the Soil Conservation Service design criteria for small dams was provided by Hurricane Agnes, one of the most ravaging storms to hit the eastern United States (U.S.D.A., 1972). The S.C.S. dams in the flooded areas held firm and saved $22 millions worth of damage for an initial cost of $38 millions.

As a comparison of four synthetic unit hydrograph methods (involving Snyder's equation, the Soil Conservation Service Method, and two others) Morgan and Johnson (1962) found peak flow predictions varying between 198 per cent and 69 per cent of the observed peak flows for 12 selected basins. The use of observed lag times improves all the methods. All methods are worse for small catchment areas than large ones.

By analysing 90 flood bydrographs on 29 British catchments, Nash obtained the parameters of the linear operator assumed to connect the effective rainfall and storm runoff. The separation technique for quick return flow was an arbitrary one based on a factor of the time to peak whilst losses were assumed at a constant rate throughout the storm. Nash's attention was consequently greatest on simple storms of short duration. Three moments, M1, M2, and M3—equivalent to a measure of lag, its variability, and the skew of the hydrograph—were derived directly from the effective rainfall and quick runoff values. They were related to catchment characteristics (see Weyman, 1975).

Painter (1971a) used a time-of-rise definition for time-to-peak in 26 catchments in north-east England. He related this variable to antecedent flow and storm rainfall parameters, but for standard values of these parameters time-to-peak is predictable from the length and slope of the main stream together with a minimum infiltration rate index for the soils in the catchment (Painter, 1971b).

In unit hydrograph analyses performed by the United Kingdom Flood Study 140 catchments were used (see Lowing and Newson, 1973), with an average of 10 flood events for each. Unit hydrographs were derived by matrix inversion followed by smoothing and the adoption of a triangular approximation to the resultant hydrograph to yield three parameters, time to peak (T_p), peak discharge (Q_p), and width-at-half-peak (T_w). Averages of these parameters for each catchment were used in multiple regression analyses. The basic prediction is that of T_p:

$$T_p = 46 \cdot 6 \ \text{MSL}^{0 \cdot 14} \ \text{S1085}^{-0 \cdot 38} \ (1 + \text{URB})^{-1 \cdot 99} \ \text{RSMD}^{-0 \cdot 40}$$

where MSL is mainstream length in kilometres, S1085 is a measure of mainstream slope, URB is the proportion of urban development in the catchment, and RSMD is a measure of excess runoff. Q_p and T_w are predictable from T_p and a better estimate of the latter can be obtained from flow records for the catchment concerned in order to get the characteristic lag time. Percentage runoff is found to depend on soil, urban, and antecedent moisture indices.

For urban areas, the design of storm-water drains and sewers demands a method of predicting hydrographs from entirely impervious surfaces such as car parks, air strips, and large roofs. In fact the urban engineering profession pioneered hydrological methods in Britain long before their application to natural catchments. Chow (1962) reviews the history of peak runoff methods as does Watkins (1951).

The rational method was partly replaced by the unit-hydrograph method in the 1930s. The most popular hydrograph method for design

of sewers in Britain is now that developed by Watkins (1962, 1963) at
the Transport and Road Research Laboratory. He analysed 286 storms on
12 catchments. A catchment at Harlow was analysed prior to urbanization
to establish the runoff pattern for upaved areas. It appeared that for the
intense summer storms, which are most important in flooding from
paved areas, natural areas are unlikely to contribute a significant portion
of the peak flood. For the other 11 urbanized areas, therefore, the
percentage of runoff from storm rainfall was calculated on the basis of
the paved area only. It was concluded that the assumption of 100 per
cent runoff for the paved areas would not lead to significant over-
design. Thus it is possible to convolute a time/area diagram of the paved
catchment with an average storm profile from the storms analysed. Thus

$$q_1 = i_1 A_1$$
$$q_2 = i_1 A_2 + i_2 A_1$$
$$q_3 = i_1 A_3 + i_2 A_2 + i_3 A_1 \ldots$$

Where $q_{1,2,3}$ are successive discharge ordinages $i_{1,\,2,\,3}$, are successive rainfall ordinates and $A_{1,\,2,\,3}$, are successive area ordinates.

It has already been mentioned that the unit hydrograph assumes that
the oversimplified input discussed above is acted on by an invariant
linear system of storages. Thus runoff is subject to storages in overland
flow, subsurface flow, and channel flow. None of these storages is linear
and their relative role in each runoff event varies. Linsley's (1967) view
is that 'it seems most improbable that any amount of mathematical
treatment can bring the concept to a level of utility consistent with
modern advances in bydrology. In the computer era we need not be
constrained by the computational difficulties which made the unit
hydrograph such an attractive tool only a few years ago.'

Nash, however, concludes 'The position with regard to hydrograph
analysis in general is scientifically entirely unsatisfactory, but as a
practical engineering tool it stands or falls by the criterion of whether it
leads to results of sufficient accuracy, or indeed whether a better
method can be obtained'.

The whole topic of alternatives to the unit-hydrograph technique and
the introduction of greater degrees of physical reality is dealt with by
Weyman (1975).

Flood routing

Information is often required on the shape of the flood hydrograph,
or its attenuation with distance, quite far down a large river system.

Because of its strict assumptions of spatially uniform rainfall the unit-hydrograph technique is seldom applied to large catchments and an alternative method must be used to design or flood warning calculations lower down. This usually consists of 'routing' the unit hydrograph for one, or a series of upstream catchments down to the point of interest. Routing consists of determining the velocity and storage characteristics of a river (and flood plain) reach between two points and using this information to assess the speed of flood wave translation down-river and the gradual attenuation of its initial peaky shape. There are two basic methods, empirical and theoretical. The former requires records of the passage of a series of flood hydrographs at the two points of interest so that, from them, parameters of the channel storage equations can be fixed. For instance the oft-used *'Muskingum' method* uses two parameters, K and x, in conjunction with instantaneous values of inflow and outflow to the reach in the following equation:

$$S = K[x.I + (1 - x)D]$$

where S equals storage, I inflow, and D outflow. Trial values of x are used (around 0·25 normally, but varying between 0 and 0·5) to adjust the value of K until a plot of $[x.I + (1 - x)D]$ versus S becomes linear and without hysteresis between rising and falling values of S. K is in fact the slope of that plot. S is known throughout from the equation:

$$\frac{I_1 + I_2}{2t} - \frac{D_1 + D_2}{2t} = S_2 - S_1$$

where t is the interval chosen for routing and equals the time between I_1 and I_2, D_1 and D_2, S_1 and S_2. The method not only requires data on streamflow but assumes that there is no hysteresis in the stage/discharge relationship for the reach and that velocity is independent of stage or discharge. These conditions are immediately violated when flow occurs across the flood plain where velocities may be halved. However, the method is simple and can be performed graphically by assuming x is zero and lagging the upstream hydrograph by successive intervals of K (which has the dimension time) derived by analysis.

The theoretical methods of routing (Price, 1973) involve either convection/diffusion equations or the *St. Venant equations* of continuity and momentum which model the one-dimensional bulk flow of water in a river. The flood wave is treated as a *kinematic wave* and the requisite data for the reach are slope, friction, inflow, cross-section, gravitational acceleration, stage, and discharge. Flood plain flows can also be included. These equations are also the basis for the theoretical modelling of runoff inputs from slopes down to the outflow of small experimental basins.

4 The flood hazard and the engineer's reaction

Calibration of hazard

Hewitt and Burton (1971), define the flood hazard as comprising:
1. loading and abrasional damage—i.e. the mechanical effects of flood waves or currents in over-bank flow,
2. drowning,
3. communications barrier effect,
4. contamination of food and water, deterioration of materials of all types,
5. housing loss,
6. disruption of social-economic activity,
7. interference with water-borne transport and spoiling of agricultural land.

Sheehan and Hewitt (1969) list floods first in a world table of natural disasters during the 20 years after 1947, comprising 209 flood disasters against 148 by the second-ranked typhoons, hurricanes, and cyclones. Floods caused 39·2 per cent of the total loss of life in natural disasters during that period but the list is not restricted to river floods. There is reason to suggest that the position of floods in the table is a result of devastation in developing equatorial countries since, in a similar table for the United States, flooding is quite low on a list of injuries and deaths caused by hazards but second only to hurricanes in the value of damage caused ($400 000 000 per annum). One contrast between the United States and Canada, where hazard studies are common, and the British Isles is that we do not experience their large range of natural hazards. For instance, in the London region of south-western Ontario Hewitt and Burton discuss the flood hazard, the hail hazard, the drought hazard, the heavy rainfall hazard, freezing rain, high winds, tornadoes, snowfall, and hurricanes! Britain is equable by comparison and does not have as large a literature or as developed a policy on flood hazard problems as North America.

The University of Chicago's Research Paper series (1942, 1957a and b, 1961, 1962a and b, 1964) indicates the length of the North American lead and Burton's offering on England and Wales (1961) rightly stresses the lack of hydrometric data and flood loss data required for sophisticated analyses here. Nevertheless, Burton concluded that without the sophisti-

cation of the North American studies, engineering work and planning
activity at that time were largely successful in preventing large-scale loss
of life and property in the two countries (Scotland and Ireland were not
treated but the same conclusion is probably valid there too). Certainly
the haphazard occupation of flood plains did not seem to be the problem it was in North America river valleys.

Recently, however, a number of teams have begun research on flood
hazard in Britain. One is based at Middlesex Polytechnic (formerly Enfield
College of Technology). Its work is described in reports by Penning-
Rowsell, Parker, *et al.* (1972 and 1973); it is intially studying the lower
Severn, centring on Gloucester. Techniques include *land-use survey* in
the area of risk and a *questionnaire survey* of the residents there. The
latter was designed to determine how well perceived the flood problem
was and what adjustments were made to it. The answers revealed that
flooding was recognized as a problem but not generally considered greater in
importance than the hazards of noise, poor services, and so on. Despite
the awareness of floods, few took steps to avoid the hazard, only
15·8 per cent insuring their property, 13·4 per cent moving valuables to
higher levels, and 9·6 per cent sandbagging. Six people (0·9 per cent)
kept a boat primarily for use in floods.

It was established that six hours' warning of floods would make
considerable savings on damage and the numbers of people affected.
Such warnings are part of the response to flood hazard higher up the
River Severn at Shrewsbury where Harding and Parker (1972) find that
over 50 per cent of residents on the flood plain do not consider there to
be a flood problem in the area. Other hazards such as traffic noise and
lack of entertainment were more in the minds of those who answered
their questionnaire. The possible reasons quoted for this situation are
the well-developed warning system which reduces losses, the recent construction of the Clywedog dam on the upper Severn, and the absence of
severe floods in the period immediately prior to the survey.

Most studies of this type conclude that information on and perception
of the flood hazard are only widely present after a recent major flood
event. This is an ideal time for the compilation of flood-damage data,
possibly by universities or water authories. An essential task is to compile
flooded-area maps, both urban and rural (Fig. 4.1). Porter (1970) was
able to construct *depth/damage curves* after the July 1968 flood near
Bristol. These curves plot the increase in damage to various types of
residential and commercial property as the depth of flooding increases
and they have now been prepared for the subdivisions of land-use de-
limited at Gloucester by the Middlesex Polytechnic project. The experi-

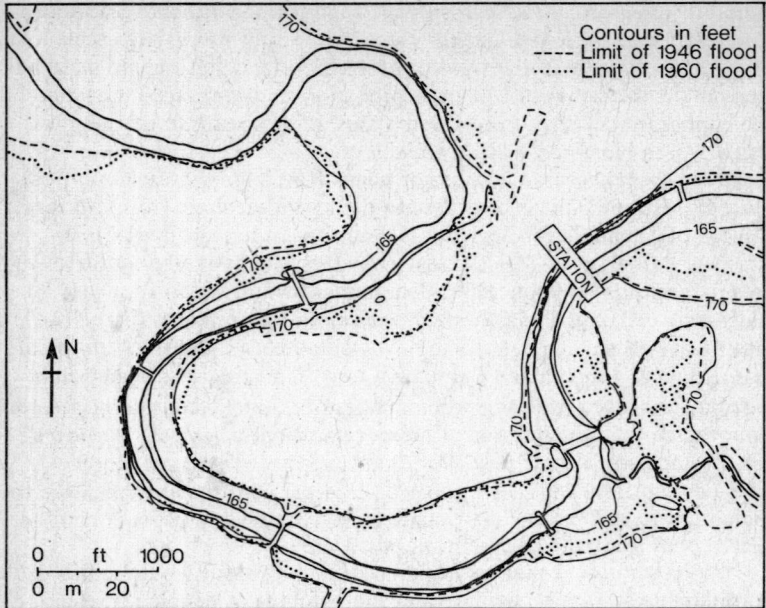

Fig. 4.1. Shrewsbury flooded-area map for 1946 and 1960 floods (after Harding in White, G. F., ed., 1975 *Natural Hazards*, O.U.P. New York.)

ence of the Task Force on Flood Plain Regulations (1962) in the U.S.A. suggests that the curves are similar for each class of enterprise or structure and therefore land-use surveys and flooded-area maps can be combined to cost the likely damages on flood plains. Parker (1973), however, mentions other complicating factors such as flood duration, velocity, response to warnings, etc. and concludes that depth/damage curves are only a first step.

Returning to the action of the individual, Sterland (1973) has sub-jected staff of the Trent River Authority to a questionnaire on their hypothetical tolerance to, and requirements of, compensation for floods of various levels entering their homes. He found that the average toler-ance for a flood which threatened life was once in 156 years and that for such an eventuality £7400 would be required (average). Within the 156-year period there would be a nested hierarchy of smaller floods and the total compensation required would be £13 975 or about £90 per annum. Such a figure, based on the probabilistic series of floods, if more reliably established, could be used as a basis for the economic consideration of hydrological schemes.

The subtleties of *cost-benefit analyses* in such situations are discussed by the Local Government Operational Research Unit in a study of Towcester (1971). One cannot count as benefits both the increased value of houses in the flooded area after protection and the costs of clearing up after the flooding since the former is a manifestation of the latter. This is called *double counting* by Chambers (1973) who draws up the personal balance-sheet of a property buyer in a flood-risk zone.

We may summarize the flood hazard situation in Britain as being characterized by episodic high levels of perception and cost data (after major floods) and by a lack of cost/benefit material which results in most schemes being cost/impact instead. Schemes are mainly local rather than comprehensive and the planning requirements for flood plains are in most cases not legally constrained. However, there are only local examples of the large-scale dangerous development of the flood plain (for example in the Lea Valley, as demonstrated by Addyman, 1973, in Fig. 4.2).

The remedies

The remainder of this chapter deals with basic methods available for flood protection with examples of their use. The choice between them may be based on their effectiveness, their impact on the local area, and the available funds. We should also include the available technology as a variable; for instnace, *structural adjustments* to floods involving the design of embankments, channels, bypasses, and so on are at present backed by more hydrological and hydraulics research than *non-structural schemes* such as *flood warning* or *watershed management* which await the thorough 'coming-of-age' of scientific hydrology before becoming widespread. More simply, the choice is between our knowledge of the peak levels and our knowledge of the complete hydrograph plus meterological conditions.

Nixon (1963) itemizes the available structural remedies for flooding used by engineers. Each is associated with flooding of a specified return period by analysis of peak flow data from a gauge in the stream at the site or close-by. As well as general maintenance of channels ('river training') the passage of a flood without damage may involve the construction of embankments, flood walls, relief channels, enlarged channels, pumping schemes, and reservoirs. The latter solution is the most expensive but has the considerable advantage that flood prevention becomes part of general water-resource planning by checking rapid or excessive runoff at or near its source in the uplands. The success of reservoirs such as the Clywedog in the upper Severn catchment has meant that *regulating reservoirs* and trans-basin aqueducts have been chosen as a key element of water-resource development in England and Wales until the next century.

The Clywedog Dam, for example, was set up by Act of Parliament in 1963. The twelve water authorities combining in the Clywedog Reservoir Joint Authority (from the Montgomeryshire Water Board down river to the Bristol Waterworks Company) were charged by the Act to maintain certain minimum dry-weather flows at Bewdley, the half-way stage on the Severn between source and estuary). These flows were fixed according to the water-supply requirements of the authorities themselves. To reduce floods downstream the reservoir has to be drawn down between May and October so that it is capable of containing the winter floods from catchments north of Plynlimon for later gradual release. On two occasions the reservoir has overflowed, but for most of its life releases are controllable by valves with the river level downstream in the town of Llanidloes being used as an index. The turbulent flood flow of the combined Dulas and Severn is in marked contrast after heavy rain to the tranquil, controlled Clywedog. The Reservoir is also used for amenities such as sailing and fishing. However, Harding and Parker (1972) consider that the beneficial influence of Clywedog is overemphasized by the flood-plain dwellers at Shrewsbury. The catchment to the Dam comprised only 2·5 per cent of that of the Severn at Shrewsbury. Thus while the Clywedog has helped the flood situation at Llanidloes and at Newtown, which was seriously affected in 1964, there are a number of important tributaries which contribute unregulated floods between Newtown and Shrewsbury. While heavy rain localized over Plynlimon may be stored by Clywedog this is seldom the case in very severe conditions (for example in August 1973) and runoff from the Berwyn Hills reaches Shrewsbury virtually unchecked (except for the Lake Vyrnwy catchment). Even at Newtown, Clywedog storage has not been used as the sole solution; the channel has been improved and flood embankments have been made part of an amenity scheme through the town.

Perhaps a prime example of the constraints on flood protection schemes for important, densely settled lowland centres is provided by that at Bath, Somerset. The Bristol Avon catchment is liable to heavy summer rainfall but the city has also been inundated during the winter, as in 1960. Of the total catchment area of the Bristol Avon (2220 km^2), 75 per cent is upstream of Bath. The city has been affected by twenty

EDMONTON & CHINGFORD – 1870

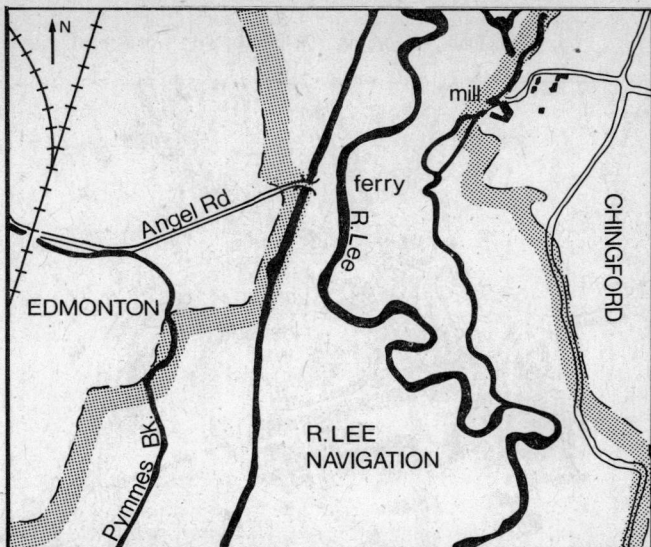

N

mill

ferry

R. Lee

Angel Rd

EDMONTON

CHINGFORD

R. LEE
NAVIGATION

Pymmes Bk.

EDMONTON & CHINGFORD – 1920

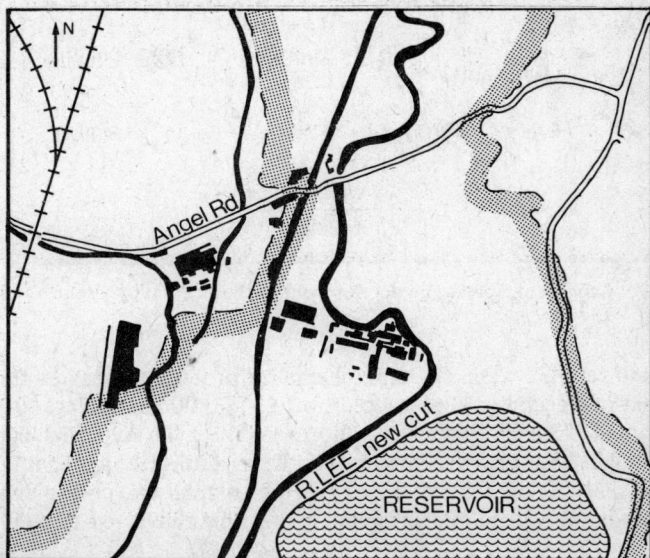

N

Angel Rd

R. LEE new cut

RESERVOIR

EDMONTON & CHINGFORD – 1968

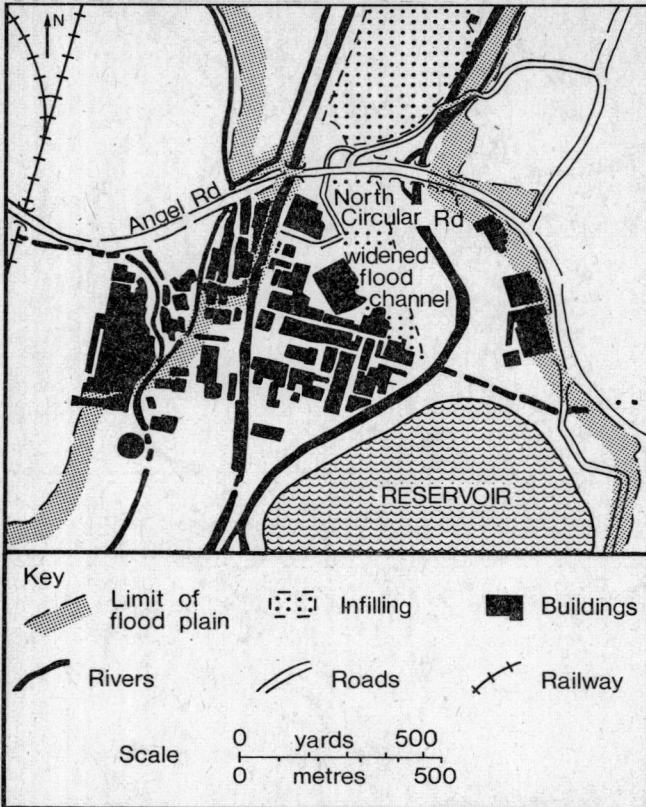

Fig. 4.2. Growth of flood-plain development in the Lea Valley (from Addyman, 1973)

floods since 1882 when the first scheme for protection was put forward and that of December 1960 caused over £1 000 000 in damage. Between 1964 and 1973 a comprehensive improvement of the Avon through Bath was brought about by the Bristol Avon River Authority under their Engineer, Mr. Frank Greenhalgh. There was virtually no choice as to the methods employed—a bypass channel would have involved tunnelling under the surrounding hills and embankments would not have stopped flooding up the sewers and culverts or through the permeable ground. The ground level could not be raised without affecting many valuable properties and a historic bridge, attractive to tourists, was also involved.

Bath is moderately well endowed with hydrological data upon which to base the design of such a scheme, with documentary records of floods back to 1823, flood marks of maximum levels on a bridge pier for over a century and a river flow gauging-station on the Avon with over 30 years of data. Bankfull discharge before improvements was 170 cumecs. The peak discharge of 1960 was measured by current meter as almost 368 cumecs, whilst those 1882, 1894, 1932, 1933, and 1947 were calculated using the Manning formula with roughness values determined by discharge measurements in the floods of 1953, 1954, and 1955. The chosen value of n was 0·03566, similar to the textbook value for a river 'in very bad order'. The two other highest floods (1882 and 1894) were also found to be between 340 and 370 cumecs.

Although such data would be adequate for small schemes, the Bath flood protection works were estimated to cost about £2 000 000 over 9 years. Consequently, to determine flood levels and discharges under various improvement strategies a 1:240 (horizontal scale) *scale model* was set up by Sir Alfred Pugsley of Bristol University. Its prototype was an 8-mile stretch of the Avon between Saltford and Bathampton. Following these investigations a ten-point programme was developed (see Fig Fig. 4.3a) involving a new bridge (Churchill Bridge), grading and removal of bends downstream of Twerton, protection at Newbridge, sluices at Twerton, new river walls through the City, a sluice at Pulteney Weir, sewer reconstruction to meet the new river level and dredging to a new river grade (see Fig. 4.3b). For design purposes the discharge of 340 to 370 cumecs was regarded as 'once a century'. Negotiations for the scheme involved central and local government (both City and County), British Rail, gas, electricity, water and other services, and, not least, riparian owners. Vertical channel banks were required in order to obtain the maximum channel capacity without the need to requisition property. The sluices nearest to the historic Pulteney Bridge were first approved as a radial gate by the Royal Fine Arts Commission and designed by a landscape architect in full liaison with the engineer. Such is the value of river scenery to the centre of Bath.

The sluices are designed to keep water levels moderately high during low flows (the function of the weirs they replace was the same—for navigation and milling purposes). As water levels rise during floods,

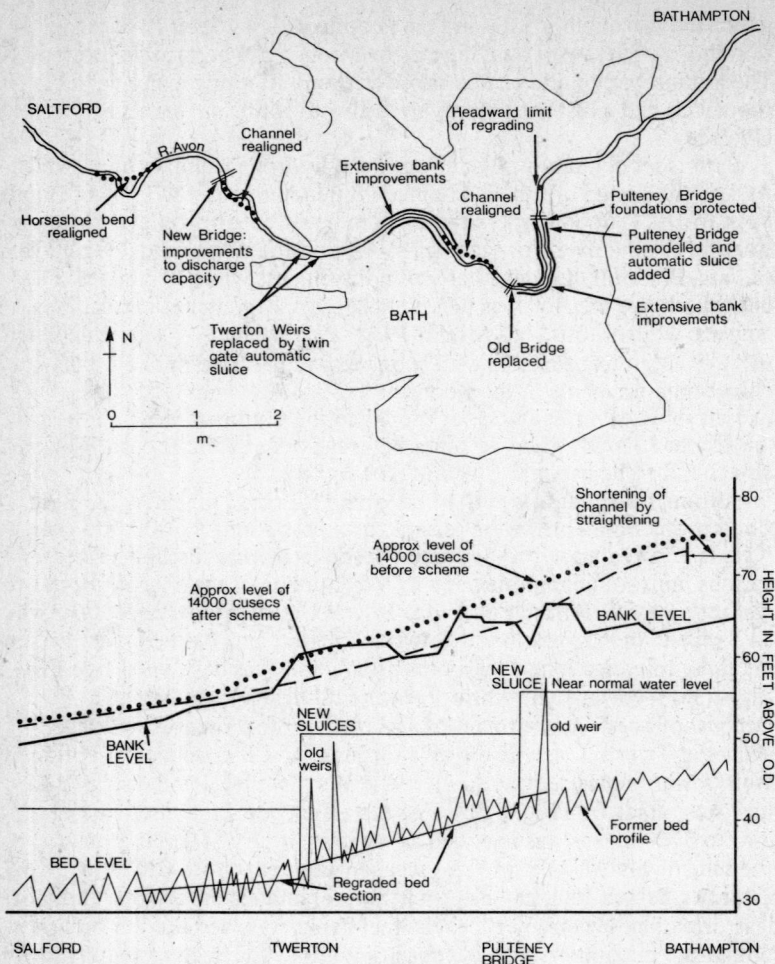

Fig. 4.3. (a) Plan of Bath's flood protection scheme
        (b) Profile of River Avon before and after improvements

however, the radial gates are counterpoised with water from the rising river so that they open and automatically prevent afflux by giving a temporarily wider channel. At Pulteney Bridge a horseshoe-shaped weir alongside the sluice-gate has been built to create a pleasing pattern of

turbulence and sound whilst the sluice-pier forms an island with trees
and shrubs. River users were also concerned about the effect of the weir
on fish movements and a fish pass was built into it. A restaurant may
eventually be incorporated in the scheme and paths, lawns, and seats are
already in use.

Harding and Parker (1972) describe the sophisticated *non-structural
approach* to the flood hazard in Shrewsbury. Although a short stretch of
protective flood bank has been constructed and planning controls have
enforced raised foundations on new building since 1947, the emphasis
is on 'flood danger warnings' perceded by 'flood alerts'. The Severn
River Authority have been able to provide 24—36 hours' warning of a
flood. The local authority and Chief Constable are warned and flags are
posted on bridges, yellow for an alert or red for a confirmed warning.
Motoring organizations post diversions. Messengers and flood wardens
warn residents while the Police contact businesses. Householders are
asked if they need help to move furniture. Thus a system of public
adjustment to flooding in Shrewsbury is superimposed upon purely
private action taken by individuals.

The basis of most flood warning schemes is the service of heavy rainfall warnings offered by the Meteorological Office. The same organization
provides estimates of the wetness of catchments in its Soil Moisture
Deficit Bulletins since antecedent conditions are critical in determining
the extent and timing of flooding. From then on it requires some idea
of the typical flood hydrograph for the catchment concerned to time
the flood's arrival downstream and its duration at sites of risk. Often a
simple graph of upstream versus downstream levels reached in previous
floods suffices for alerts downstream after the flood has passed the
upstream station.

Inevitably, improvements in remote-sensing will allow more effective
on-line data collection on rainfall and river levels and processing to
achieve warnings semi-automatically. Perhaps the greatest chance of
advance is, however, in meteorology and *synoptic climatology* where
calibration of the conditions leading to heavy rainfall can be achieved.
Thomas (1960) put down some early predictive work on *depression
tracks* and the incidence of cyclonic rainfall; he concluded that warm
sector rain was most subject to orographic exaggeration. More recently,
Browning *et al.* (1973) have made intensive investigations of rainfall
patterns within depressions. Holgate (1973) attempts a rainfall fore-
cating method based on synoptic conditions for the Lake District and
North Wales where intensities of 6 mm per hour appear critical for
flooding if mainained for 6 hours over the short, steep catchments. In

these areas the arrival of fronts is crucial for heavy rain, cold fronts being most likely to give such falls. Upper air information is also required on humidity and wind speeds. Of 144 heavy falls observed in the two areas 108 were correctly forecast but, important for credibility, 216 falls were actually forecast in total. Lowndes (1968, 1969) divides his study of synoptic conditions for heavy daily falls in North Wales into summer and winter periods. His results are similar to those of Holgate in that a partly occluded depression, wave depression, or wave is the basic cause. However, there is a total of seven criteria and these are subject to slight seasonal variation. Matthews (1972) discovers a significant relationship between rainfall intensity and synoptic type in the Midlands and North Wales. Barrett (1973) forecasts daily rainfall totals on the basis of an index derived from satellite photographs of cloud patterns and conventional weather data for Valentia in south-west Ireland. Rainfall for the day ahead is categorized as 'no rain', 'light rain', 'moderate rain', or 'heavy rain'. Over a test period 74·3 per cent of forecasts were in the right category of fall or gave the correct choice of adjacent categories. Without 'either/or' forecasts the method attained 50 per cent accuracy.

Neither Holgate not Barrett claims to deal with convectional rainfall situations, although Lowndes and Matthews include thundery situations. Saunders (1966) tests eight methods of forecasting thunderstorms based on lapse rates and upper air readings of temperature and dew-point depression. He concludes that over most of the East Coast the subjective impression of the forecaster was superior (achieving 79 per cent accuracy). Certainly the spatial and diurnal incidence of thunder shows usable regularities but the movement of convective storms which can be detected by radar (Newton and Frankhauser, 1961) is likely to be more important. New activity tends to develop on the right-hand side of the squall line relative to the mean wind. This is consistent with physical considerations and with supply and demand requirements of the storm-water budget. In Britain, the *Dee Weather Radar Project* (Harding, 1972, Harrold *et al.*, 1973) has discovered a suitably strong relationship between surface rainfall rates and radar reflectivity to allow use of radar in operating the reservoir systems in the Dee catchment.

5   Floods as a part of the dynamic physical system

It would be a misrepresentation to leave the impression with the reader that the flood hazard represents a basically 'inefficient' natural system which involves rivers regularly overtopping their banks to create havoc. We have to examine the role of the river channel and flood plain in the absence of man's activities, determine the nature of changes in such a wholly natural system through time (in response to climatic changes), and decide how the activities of man are likely to alter the system.

Flooding and the physiographic stystem

The graphic effects of a large flood on the landscape have often led geomorphologists to incline once more towards the pre-Huttonian *'catastrophist'* view of the landscape which sees floods as infrequent, highly active spells between long periods of inactivity. However, whilst flood erosion gives the geomorphologist the chance to investigate processes of landform development concentrated in time, it is clear from the results of joint hydrological and erosional studies that events of less magnitude can achieve more effect because of their greater frequency. Naturally, the precise type of action is important: for slope failure or bedrock erosion in river channels there may be thresholds of action which are not crossed until very high flows are reached, whilst many aspects of channel morphology are clearly related to moderate flows. It may happen that floods of very low probability leave a lasting effect on the subsequent balance and location of erosive processes. Much of the spectacular action witnessed after big floods results from flow in ephemeral channels such as gullies; thus, only on infrequent occasions is flow experienced in those locations and the intervening period of weathering tends to provide quantities of material for movement. The association of several indices of river channel morphometry with the bankfull discharge confirms the long-term effect of more moderate flows. *Bankfull* (or *'channel-forming'*) *discharge* occurs on average 0·6 per cent of the time, or about 2·2 times per year in Britain according to Nixon (1959), whilst in the United States it has been found that rivers equal or exceed bankfull two out of every three years. At bankfull, the surface slope of the stream becomes uniform, the whole wetted perimeter is being washed, and bed forms are generally in motion. As a consequence, sediment transport is maximized (overbank flow only leads to deposition in the slacker water) and it is not surprising that both plan features (meander wavelengths) and profile features (riffle spacings) show correlation with bankfull discharge.

The corollary of the above remarks is that the *flood plain* itself is seldom the site of present-day geomorphological action by rivers except in so far as they both erode and deposit flood-plain material laterally by migration of natural channel sinuosities. Thus the flood plain does not represent the necessary extra channel capacity eroded by overbank flows and indeed it may even be wider than predictable from the migrations of the present-day stream. Today's streams are frequently 'misfits' (Dury, 1958) following a decline of precipitation or the end of glacial meltwater runoff in the Post-glacial. Wolman and Miller (1960) sum up the magnitude and frequency of geomorphological processes diagramatically (see their Fig. 1, p. 56) and metaphorically in a fable which tells of the superior work done by the steady man to that achieved by a dwarf who never rests and a giant who mostly sleeps!

Flooding and climatic change

The essential characteristics of rainfall leading to flooding in Britain have already been outlined in Chapter 3. Obviously any climatic trend which affected heavy frontal falls in the west or convective storms over central and eastern Britain would be of direct relevance to flooding. Lamb (1972) has pointed out a decline in westerly weather types over the British Isles since 1955 and a growth of 'blocking' anticyclonic conditions. A similar situation developed during the last quarter of the nineteenth century. Lamb's work does not refer to the effects of such changes on hydrological systems but it is clear that anticyclonic weather is likely to be drier than cyclonic except in summer when the col situation can produce thunderstorms. Bleasdale (1970) says that the changed pattern leads to 'increased risk of quasi-stationary situations more favourable for the random occurrence of unusual rainfall events'. He distinguishes two periods of unusually heavy falls of rain, 1912–24 and 1952–68; whilst the latter corresponds to Lamb's era of declining westerlies, the former is wid-way through the period of prominent westerlies!

Howe, Slaymaker, and Harding (1967) analysed daily falls of over 63·5 mm for Lake Vyrnwy and the Elan Valley. Both showed a marked increase in the period 1940–64 compared with 1911–40, yet both are stations which receive mainly frontal rain from a westerly circulation (supposedly in decline during the latter half of the period). However, since no trend is discovered in the annual rainfall for the stations it appears that the effect is one of increasing rainfall intensities—corroborating Bleasdale's views. Finch (1972) finds a marked upward trend in the frequency of exceptional falls (see Chapter 2 for definition) at Effingham in Surrey between 1968 and 1971 but this is a short record on which to base trends and there are two long periods with no exceptional falls. Rodda (1970) plots the rainfall frequency (daily totals) for Oxford, taking four discrete periods of record (see Fig. 5.1). The increased frequency of heavy rain can be gauged from the fact that for the period 1931—45 (23 storms) the return period of a fall of 63·5 mm was approximately 50 years whereas for 1941—65 (28 storms) it was reduced to approximately 10 years. The growth of urban development may itself affect rainfall. Atkinson (1968) has shown that thundery rainfall during the summer months shows distinctly higher values over London (see Fig. 5.2). Contributions to this effect may come from increased roughness over the urban area, the role of pollutants in providing condensation nuclei, higher vapour pressure in the city, and more humid thermals.

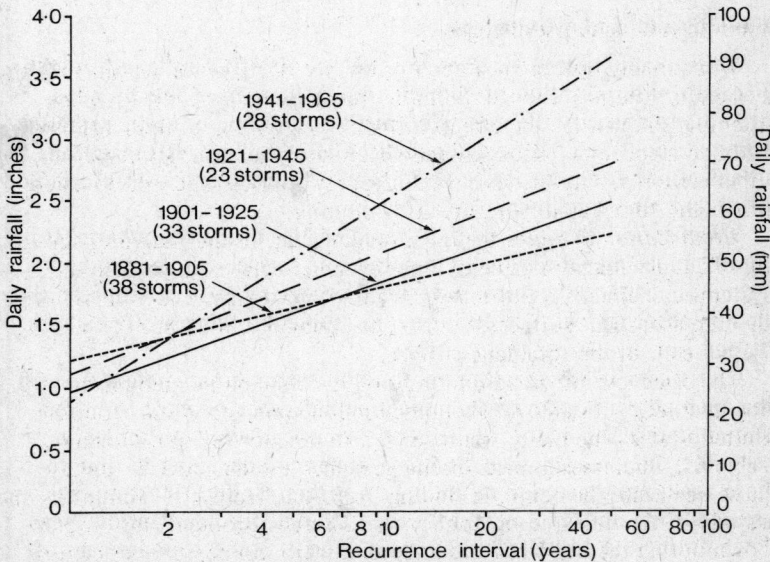

Fig. 5.1. Frequency of daily falls, Oxford (from Rodda, 1970)

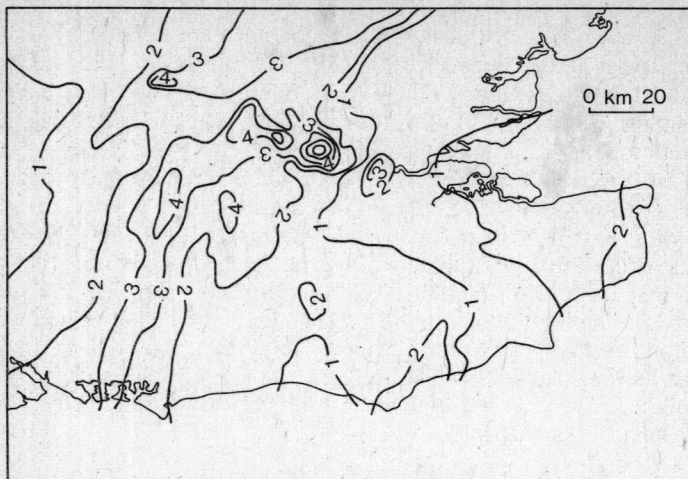

Fig. 5.2. Thunder rainfall (inches) in warm front situations, summer, 1951–60
(from Atkinson, 1969)

Flooding and land-use changes

Whilst many hydrological records are not of sufficient length to allow
successful interpretation of climatic trends, there have already been
attempts to quantify the effects of man's occupation of drainage basins
from data collected before and and after land-drainage, afforestation, or
urbanization. Comparisons between nearby forested and non-forested or
urban and rural catchments are also common.

Urbanization provides the most rapid change in the characteristics of
runoff, replacing permeable by impermeable surfaces and a natural
system of channels by storm sewers and other drains. It is estimated that
the non-absorbent surface of Surrey has trebled in the last 50 years to
7½ per cent of the total land surface.

The effects of urbanization on flood hydrograph parameters are well
documented, with an overwhelming preponderance of work from the
United States. The major results of the American work are shown in
Table 6.1—increases in peak discharge values and decreases in time to
those peaks are the common finding. In Britain Hollis (1974) reports on
a study of the unit hydrograph for the Cannons Brook at·Harlow New
Town during the period of urban development. Some 16·6 per cent of
the basin became covered by impervious surfaces linked by drains to the
channel between October 1953 and September 1968. The peak of the

TABLE 6.1

A summary of the restults of studies of urban and suburban influences on flood flows in the United States

Author/date	State/city	Findings
B. L. Bigwood and M. P. Thomas, 1955 (U.S.G.S. Circular 365)	Connecticut	Flood peak formula has coefficient of 0·8 for the whole state but 3·0 for urban/residential areas (increase of 3·8).
H. P. Ramey, 1959 (*Jnl. Hyd. Div. Proc. A.S.C.E.*)	Chicago	Runoff increased by 3 times in last 30 years.
A. L. Tholin and C. J. Kiefer (1960) (*Trans. Am. Soc. Civ. Engrs.*)	Chicago	Commerical/industrial areas have peak flows 3·5 to 4·0 times those of residential areas.
S. W. Wiitala, 1961 (U.S.G.S. Open file rept.)	Michigan	Peak flows increased by 3 times; lag times reduced by 70 per cent.
R. W. Carter, 1961 (U.S.A.S. Prof. Paper 424–B)	Washington, D.C., Maryland, Virginia	Flood peaks of all return periods increased 1·8 times. Coefficient in lag-time prediction reduced from 3·1 for rural areas to 1·2 for urban (lag reduced by 80 per cent).
A. O. Waananen, 1961 (U.S.G.S. Prof. Paper 424–C)	New York State, Connecticut	Peak flows increased nearly 3 times.
D. G. Anderson, 1963 (U.S.G.S. Prof. Paper 475–A)	Northern Virginia	Peak flows increased by 2 to 8 times; lag times reduced by 85 per cent.
J. R. Crippen, 1965 (U.S.G.S. Prof. Paper 525–D)	California	Peak runoff increased by 1·4 times; lag-time prediction coefficient reduced from 3·7 to 2·6 (reduction of lag times by 30 per cent).
W. H. Espey, C. W. Morgan, and F. D. Masch, 1965 (Tech. Rept. Texas Water Commission)	Texas	Peak flows up by half; time of rise reduced by 46 per cent.
K. V. Wilson, 1967 (U.S.G.S. Prof. Paper 575–D)	Mississippi	Mean annual flood between 2·5 and 4·5 times rural although reduced effect at higher return periods due to sewer storage.
L. A. Martens, 1968 (U.S.G.S. Water supply Paper 1591–C)	North Carolina	The 20-year flood is doubled by urbanization. However, the effect becomes negligible above the 50-year return period.
C. E. Seaburn, 1969 (U.S.G.S. Prof. Paper 627–B)	Long Is., New York	Runoff increased by 270 per cent; peak discharge increased by 2·5 times; width of hydrograph at half-peak reduced by 62 per cent.

unit hydrograph increased by 4·6 times whilst the time of rise was more than halved; however, these results are largely biased to moderate rather than large flood events and to summer rather than winter floods. For winter floods and floods with peaks over 4 cumecs (20-year return period) the effect of urbanization was minimal, possible because the catchment was naturally fairly impermeable (London clay with boulder clay) or because the surface-water drains' capacity is exceeded by discharges of a return period greater than one year with consequent delays in runoff. Hollis makes the point that the density of the artificial drainage system is only of the same order as the fully expanded natural channel network during floods (see Hanwell and Newson, 1970, Fig. 15) and the latter is not roofed. However, the natural channel system may take some time to extend and it is not surprising that the most obvious effect of urban drainage is on the timing of floods. Gregory (1974) finds lag times are under half pre-urban values in his Exeter catchment Only slightly more runoff produces a peak flow two or three times larger than it was before urbanization.

Regarding the use of the urban fraction of catchments for regression analysis of floods data, the major problem is that a wide variety of densities of development and impermeability factors is concealed by the 'grey' areas representing towns on most maps. The United Kingdom Flood Study has used the percentage of 'grey' area shown on the 1:250 000 Ordnance Survey map as a variable and it proves a significant variable in equations for unit hydrograph time-to-peak.

The upstream cause/downstream effect controversy has also raged at times over both *agricultural drainage* and *afforestation*. The effect of afforestation on floods is generally assumed to be a beneficial one although, as pointed out in Discussion of Dobbie and Wolf (1953), this may only be the case in minor floods or in isolated storms: continuous heavy rain exceeds the extra storage capacity provided by the trees.

In the early stages of site preparation for planting forests in the uplands a considerable amount of deep ploughing and drainage is done and Howe, Slaymaker, and Harding (1967) suggest that this increase of natural drainage densities has contributed to the increased frequency of flooding in the Wye and Severn catchments between 1940 and 1964. However, forest drainage is seldom noticeably effective for more than a month or so whilst the boggy areas of slope 'dewater', after which it is probably only in the larger floods that flow occurs in them. The lowering of water levels between drains clearly gives an extra infiltration capacity. Research is in progress by the Forestry Commision, by the Institute of Hydrology at the Coalburn catchment in the recently ditched Kielder Forest, and at Plynlimon in the Hafren Forest. The presence of peat is clearly important (Conway and Millar, 1960), and soil types may also vary in their reaction to both forest and agricultural drainage practices.

Obviously, both climatic and land-use changes could be taking us into an era of increased flooding, whilst man's need for space on flood plains may be taking him closer to this hazard. However, despite the possible rapidity of the change, the scientific conclusion so far must be that 'it is too early to say conclusively' and the designers of improved water-resource systems have already begun to couple the demand of a drought period with the vast supply of a flood, thus providing an artificially beneficial balance to natural extremes.

References

N.B. Some references in the text are not listed here but can be found in Table 1.1.

Addyman, O. T. 1973 *Some questions arising from flood plain problems in the River Lee catchment.* Conference of River Authority Engineers, Cranfield.

Allard, W., Glasspoole, J., and Wolf, P. O. 1960 'Floods in the British Isles'. *Proc. Instn. civ. Engrs.* **15**, 119–44.

Anon, 1849 *Memorials of the flood in the rivers of Northumberland and Durham.* Richardson, Newcastle.

Atkinson, B. W. 1969 'A further examination of the urban maximum of thunder rainfall in London, 1951–60', *Trans. Inst. Br. Geogr.* **48**, 97–119.

Barnes, H. H. 1967 *Roughness characteristics of natural channels.* United States Geological Survey, Water Supply Paper, 1849, 213 pp.

Barrett, E. C. 1973 'Forecasting daily rainfall from satellite data', *Monthly Weather Review,* **101** (3), 215–22.

Benson, M. A. 1960 'Characteristics of frequency curves based on a theoretical 1000-year record', in *Flood Frequency Analyses* (by T. Dalrymple). United States Geological Survey, Water Supply Paper, 1543–A.

—— 1962 *Factors influencing the occurrence of floods in a humid region of diverse terrain.* United States Geological Survey, Water Supply Paper, 1580–B.

—— 1964 *Factors affecting the occurrence of floods in the south-west.* United States Geological Survey, Water Supply Paper, 1580–D.

Bilham, E. G. 1935 'Classification of heavy falls in short periods', *British Rainfall,* 262–80.

Binnie, A. M. and Mansell-Moullin, M. 1966 'The estimated probable storm and flood on the Jhelum River—a tributary of the Indus', in *River Flood Hydrology,* Instn. civ. Engrs., 189–210.

Biswas, A. K., and Fleming, G. 1966 'Floods in Scotland: magnitude and frequency', *Water and Water Engrng.* **70** (844), 246–52.

Bleasdale, A. 1963 'The distribution of exceptionally heavy falls of rain in the United Kingdom, 1863–1960', *Jnl. Instn. Wat. Engnrs.* **17**, 45–55.

—— 1970 'The rainfall of 14th and 15th September 1968 in comparison with previous exceptional rainfall in the United Kingdom', *Jnl. Instn. Wat. Engnrs.* **24**, 181–9.

Bransby-Williams, G. 1922 'Flood discharge and the dimension of spillways in India', *Engineer,* **134**, 321–2.

British Standards Institution 1970 *Liquid flow in open channels: slope area method of estimation.* BS 3680, Part 5, 13 pp.

Brooks, C. E. P., and Glasspoole, J. 1928 *British Floods and droughts.* Benn, London.

Browning, K. A. *et al.* 1973 'The structure of rainbands within a mid-latitude depression'. *Quart. J. Roy. Met. Soc.* **99** (420), 215—31.

Burton, I. 1961 'Some aspects of flood loss reduction in England and Wales', in White (ed.), *Papers on Flood Problems.* University of Chicago, Research Paper Series, **70**, 203—21.

Butler, R. M. J. 1972 'Water as an unwanted commodity: some aspects of flood alleviation', *Jnl. Inst. Wat. Engrs.* **26** (6), 311—32.

Cambers, D. N. 1973 'Economic aspects of flood alleviation', in *Proceedings of a symposium on economic aspects of floods,* Middlesex Polytechnic, 33—5.

Chapman, E. J. K., and Buchanan, R. W. 1966 'Frequency of floods of "normal maximum" intensity in upland areas of Great Britain', in *River Flood Hydrology.* Instn. civ. Engnrs., London, 65—86.

Chow, V. T. 1962 'Hydrologic determination of waterway areas for the design of drainage structures in small drainage basins', University of Illinois, *Engineering Experimental Station Bulletin* **59** (65), 462.

Cole, G. 1966 'An application of the regional analysis of flood flows' in *River Flood Hydrology,* Instn. civ. Engnrs., 39—57.

Conway, V. M., and Miller, A. 1960 The hydrology of some small peat-covered catchments in the Northern Pennines. *Jnl. Instn. Wat. Engnrs.* **14**, 415—424.

Crossley, A. F., and Lofthouse, N. 1964 The distribution of severe thunderstorms over Great Britain. *Weather*, **19**, 172—177.

Cunnane, C. 1973 A particular comparison of annual maxima and partial duration series methods of flood frequency prediction. *Jnl. Hydrol.,* **18**, 257—271.

Dalrymple, T. 1956 Measuring floods. *Int. Assn. Sci. Hydrol.,* Pubn. **42**, 380—404.

Dobble, C. H., and Wolf, P. O. 1953 'The Lynmouth flood of August 1952', *Proc. Instn. civ. Engrs.* **2**, 522—88.

Dooge, J. C. I. 1959 'A general theory of the unit hydrograph', *Jnl. Geophys. Res.* **64** (2), 241—56.

Dury, G. H. 1958 'Tests of a general theory of misfit streams', *Trans. Inst. Br. Geogrs,* **25**, 105—18.

—— 1959 'Analysis of regional flood frequency on the Nene and Great Ouse', *Geog. Jnl,* **125** (2), 223—9.

Finch, C. R. 1972 'Some heavy rainfalls in Great Britain, 1956—1971', *Weather,* **27** (9), 364—77.

Fisher, R. A., and Tippett, L. H. C. 1928 'Limiting forms of the frequency distribution of the smallest and largest member of a sample', *Proc. Cambridge Phil. Soc.* **24**, 180—90.

Gregory, K. J. 1974 'Streamflow and building activity', *Fluvial Processes in Instrumented Watersheds*. Inst. Br. Geogrs. Spec. Pub. 6, 107—22.

Grindley, J. 1967 'The estimation of soil moisture deficits', *Met. Mag.* **96** (1137), 97—108.

Gumbel, E. J. 1941 'The return period of flood flows', *Ann. Math. Stat.* **12** (2), 163—90.

—— 1955 'The calculated risk in flood control', *Applied Science Research*, A(5), 273—80.

—— 1958 'Statistical theory of floods and droughts', *Jnl. Instn. wat. Engnrs.* **12**, 157—84.

Hanwell, J. D., and Newson, M. D. 1970 *The great storms and floods of July 1968 on Mendip*. Wessex Cave Club Occl. Pubn. 1 (2).

Harding, D. M., and Parker, D. J. 1972 *A study of the flood hazard at Shrewsbury, United Kingdom*. 22nd. Int. Geogr. Cong. Calgary.

Harding, J. 1972 'Rainfall measurement by radar—Dee Weather Radar Project', *World Met. Orgn. Bulletin*, **21** (2), 90—4.

Harrold, T. W., English, E. J., and Nicholass, C. A. 1973 'The Dee Weather Radar Project: the measurement of area precipitation using radar', *Weather*, **29**(8), 332—8.

Hazen, A. 1930 *Flood flows: a study of frequencies and magnitudes*. John Wiley, 199 pp.

Herschfield, D. M., and Kohler, M. A. 1960 'An empirical appraisal of the Gumbel extreme-value procedure', *Jnl. Geophys. Res.* **65** (6), 1737—46.

Hewitt, K., and Burton, I. 1971 *The hazardousness of a place. A regional ecology of damaging events*. Univ. of Toronto, Dept. of Geog. Research Pubns., 154 pp.

Holgate, H. T. D. 1973 'Rainfall forecasting for river authorities', *Met. Mag.* **102** (1207), 33—47.

Holland, D. J. 1967 'The Cardington rainfall experiment', *Met. Mag.* **96**, 193—202.

—— 1964 (1968) *Rain intensity frequency relations in Britain*. U.K. Meteorological Office, Hydrological Memoranda, 33 (appendix).

Hollis, G. E. 1974 'The effect of urbanization on floods in the Canon's Brook, Harlow, Essex', in: *Fluvial Processes in Instrumented Watersheds*. Inst. Br. Geogrs. Special Pubn. 6, 123—39.

Howe, G. M., Slaymaker, H. O., and Harding, D. M. 1967 'Some aspects of the flood hydrology of the upper catchments of the Severn and Wye', *Trans. Inst. Br. Geogrs.* **41**, 33—58.

Institution of Civil Engineers Committee on Floods 1933 (1960) *Floods in relation to reservoir practice*. London.

Kuichling, E. 1889 'The relation between the rainfall and the discharge of sewers in populous districts', *Trans. Am. Soc. civ. Engrs.* **20** (1), 1—60.

Lamb, H. H. 1972 *British Isles weather types and a register of the daily sequence of circulation patterns, 1861—1971*. U.K. Meteorological Office, Geophysical Memoir 116, 85 pp.

Linsley, R. K. 1967 'The relation between rainfall and runoff', *Jnl. Hydrol.* 5, 297–311.

Lloyd-Davis, D. E. 1906 'The elimination of storm water from sewerage systems', *Proc. Instn. Civ. Engrs.* 164, 41–67.

Local Govt. Operational Research Unit 1971 *Cost benefit analysis of Towcester flood relief scheme.* Rept. T33.

Lowing, M. J., and Newson, M. D. 1973 'Flood event data collation', *Water and Water Engrng.* 77, 91–5.

Lowndes, C. A. S. 1968 'Forecasting large 24-hour rainfall totals in the Dee and Clwyd River Authority area, September–February', *Met. Mag.* 97, 226–35.

–– 1969 'Forecasting large 24-hour rainfall totals in the Dee and Clwyd River Authority area, March–August', *Met. Mag.* 98, 325–40.

McGuiness, J. L., and Brakenseik, D. L. 1964 *Simplified techniques for fitting frequency distributions to hydrologic data.* Agricultural Handbook 259, Agric. Res. Serv., U.S. Dept. Agric.

Matthews, R. P. 1972 'Variation of precipitation intensity with synoptic type over the Midlands', *Weather,* 27 (2), 63–72.

Morgan, H. D. 1966 'Estimation of design floods in Scotland and Wales', in *River Flood Hydrology*, Instn. civ. Engrs. 59–64.

Morgan, P. E., and Johnson, S. M. 1962 'Analysis of synthetic unit-graph methods, *Jnl. Hyd. Div., Proc. Am. Soc. civ. Engrs.* 88, HY5, 199-220.

Nash, J. E. 1957 'The form of instantaneous unit hydrograph', *Int. Assn. Sci. Hydrol. Pubr* 45 (3), 114–21.

–– 1958 'Determining runoff from rainfall', *Proc. Instn. civ. Engrs.* 10, 163–84.

–– 1959 'Systematic determination of unit hydrograph parameters', *Jnl. Geophys. Res.* 64, 111–5.

–– 1960 'A unit hydrograph study with particular reference to British catchments', *Proc. Instn. civ. Engrs.* 17, 249–82.

–– and Shaw, B. L. 1966 'Flood frequency as a function of catchment characteristics', in *River Flood Hydrology,* Instn. civ. Engrs., 115–35.

Newton, C. W., and Frankhauser, J. C. 1961 'On the movements of convective storms with emphasis on size discrimination in relation to water-budget requirements', *J. Appl. Met.* 3 (6), 651–68.

Nixon, M. 1963 'Flood regulation and river training in England and Wales', in *Conservation of Water Resources*, Instn. civ. Engrs., 137–50.

–– 1959 'A study of the bankfull discharges of rivers in England and Wales', *Proc. Instn. civ. Engrs.* 12, 157–74.

O'Donnell, T. 1960 'Instantaneous unit hydrograph derivation by harmonic analysis', *Int. Assn. Sci. Hydrol. Pubn.* 51, 546–57.

Overton, D. E. 1968 'A least squares hydrograph analysis of complex storms on small agricultural watersheds', *Wat. Res. Res.* 4 (5), 955–63.

Painter, R. B. 1971a 'The hydrograph time-to-peak method of flood prediction', *Water and Water Engrng.* 75 (904), 235–7.

Painter, R. B. 1971b 'A hydrological classification of the soils in England and Wales', *Proc. Instn. civ. Engrs.* **48**, Tech Note 29, 93–5.

Parker, D. J. 1973 'The assessment of flood damages', in *Proc. Symp. on econ. aspects of floods*, Middlesex Polytechnic, 9–20.

—— and Penning-Rowsell, E. C. 1973 *Problems and methods of flood damage assessment.* Middlesex Poly. Flood Hazard Research Project, Rept. No. 3.

Penning-Roswell, E. C. 1972 *'Enfield College Flood Hazard Research Project.* Middlesex Poly. Flood Hazard Research Project. Rept. No. 1.

—— and Underwood, L. 1972 *Flood hazard and flood-plain management: survey of existing studies.* Middlesex Poly. Flood Hazard Research Project, Rept. No. 2.

—— and Parker, D. J. 1973 *The control of flood plain development: a preliminary analysis.* Middlesex Poly. Flood Hazard Research Project, Rept. No. 4.

Porter, E. A. 1970 'Assessment of flood risk for land use planning and property insurance'. Unpub. Ph.D. Thesis, Univ. of Cambridge.

Powell, R. W. 1943 'A simple method of estimating flood frequencies', *Civil Eng.* **13** (2), 105–6.

Price, R. K. 1973 'Flood routing methods for British rivers', Hydraulics Research Station, Rept. INT–111, or *Proc. Instn. Civ. Engrs.* **55** (2), 913–30.

Reich, B. M. 1962 'Soil Conservation Service design hydrographs', *The Civil Engineer in South Africa,* **4**, 77–87.

Riggs, H. C. 1968 'Frequency curves', in *Techniques of Water Resources Investigations of the United States' Geological Survey*, Chapter A–2, 15 pp.

Rodda, J. C. 1966 'A study of the magnitude, frequency and distribution of intense rainfall in the United Kingdom', *British Rainfall*, 204–15.

—— 1967 'A countrywide study of intense rainfall for the United Kingdom', *Jnl. Hydrol.* **5**, 58–69.

—— 1969 'The significance of characteristics of basin rainfall and morphometry in a study of floods in the United Kingdom', *Int. Assn. Sci. Hydrol. Pubn.* **85**, 834–45.

—— 1970 'Rainfall excess in the United Kingdom', *Trans. Inst. Br. Geogrs.*, **49**, 49–59.

Saunders, W. E. 1966 'Tests of thunderstorm forecasting techniques', *Met. Mag.* **95**, 204–10.

Sheehan, L., and Hewitt, K. 1969 *A pilot survey of global natural disasters of the past twenty years.* Natural Hazard Research Working Paper, No. 11, Toronto.

Sherman, L. K. 1932 'Streamflow from rainfall by the unitgraph method', *Engrng. News Record,* **108**, 501–5.

Snyder, F. F. 1938 'Synthetic unit graphs', *Trans. Amer. Geophys. Un.*, Part 1, 447–54.

Snyder, W. M. 1955 'Hydrograph analysis by method of least squares', *Proc. Am. Soc. civ. Engrs.* 81, Paper 793, 1–25.

Sterland, F. K. 1973 'An evaluation of personal annoyance caused by flooding', in *Proceedings of symposium on economic aspects of floods,* Middlesex Polytechnic, 21–32.

Task Force on Flood Plain Regulation 1962 'Guide for the development of flood plain regulations', *Jnl. Hyd. Div., Proc. Am. Soc. Civ. Engrs.,* HY5, 73–119.

Thomas, T. M. 1960 'Precipitation within the British Isles in relation to depression tracks'. *Weather,* 15, 360–73.

United States Dept. of Agriculture (Soil Conservation Service) 1957 *Hydrology:* National Engineering Handbook Section 4, Suppl. A.

–– 1972 'When Hurricane Agnes reigned and rained', *Soil Conservation,* 38 (2), 28–32.

United States Water Resources Council 1967 *A uniform technique for determining flood flow frequencies.* Bull. 15, Washington, D.C., 15 pp.

University of Chicago Dept. of Geography Research Paper Series

White, G. F. 1942 *Human adjustment to floods* (Res. Pap. 29).

White, G. F. *et al.* 1957a *Changes in urban occupance of flood plains in the United States* (Res. Pap. 57)

Murphy, F. C. 1957b *Regulating flood plain development* (Res. Pap. 00).

White, G. F. 1961 *Papers on flood problems* (Res. Pap. 70).

Burton, I. 1962a *Types of agricultural occupance of flood plains in the United States* (Res. Pap. 00).

Kates, R. W. 1962b *Hazard and choice perception in flood plain management* (Res. Pap. 78).

White, G. F. 1964 *Choice of adjustment to floods* (Res. Pap. 93).

Water Resources Board and Scottish Development Department 1971 *The Surface Water Yearbook of Great Britain* (1965–6) with Supplement 1965. H.M.S.O., London.

Watkins, L. H. 1951 'Surface water drainage–a review of past research', *J. Instn. mun. Engrs.* 78, 303–20.

–– 1962 *The design or urban sewer systems.* D.S.I.R., Road Res. Tech. Paper No. 55. H.M.S.O., London,

–– 1963 'Research on surface-water drainage', *Proc. Instn. civ. Engrs.* 24, 305–30 (Discussion 1964, 27, 588–612).

Weyman, D. R. 1975 *Runoff processes and streamflow modelling.* Oxford University Press.

Williams, H. Bowen 1957 'Flooding characteristics of the River Swale'. Unpub. Ph.D. Thesis, Dept. Civ. Engrng., Univ. of Leeds.

Wilson, E. M. 1969 *Engineering Hydrology.* Macmillan, 182 pp.

Wolman, M. G., and Miller, J. P. 1960 'Magnitude and frequency of forces in gemorphic processes', *Journ. Geol.* 68 (1), 54–74.

World Meteorological Organization 1973 *Manual for estimation of probable maximum precipitation.* Operational Hydrology Rept. 1.

Index